Topological Transformation Groups

Deane Montgomery
Leo Zippin

Dover Publications, Inc.
Mineola, New York

Bibliographical Note

This Dover edition, first published in 2018, is an unabridged republication of the 1974 edition published by the Robert E. Krieger Publishing Company, Huntington, New York, of the work originally published in 1955 by Interscience Publishers, Inc., New York, as part of the "Interscience Tracts in Pure and Applied Mathematics" series.

Library of Congress Cataloging-in-Publication Data

Names: Montgomery, Deane, 1909-1992, author. | Zippin, Leo, author.
Title: Topological transformation groups / Deane Montgomery, Leo Zippin.
Description: Dover edition. | Mineola, New York : Dover Publications, Inc.,
    2018. | Originally published: New York : Interscience Publishers, Inc.,
    1955. An unabridged republication of: Huntington, New York : Robert E.
    Krieger Publishing Company, 1974. | Includes bibliographical references.
Identifiers: LCCN 2017060626 | ISBN 9780486824499 | ISBN 0486824497
Subjects: LCSH: Topological transformation groups. | Topological groups. |
    Transformation groups. | Group theory. | Transformations (Mathematics)
Classification: LCC QA613.7 .M66 2018 | DDC 514—dc23
LC record available at https://lccn.loc.gov/2017060626

Manufactured in the United States by LSC Communications
82449701   2018
www.doverpublications.com

PREFACE

The last few years have seen a considerable increase in our knowledge of the structure of locally compact groups, and in this book we shall try to make some of this information conveniently available. We have a double purpose; one, to report fully on those recent developments which tend to round off the subject of the relation of locally compact groups to Lie groups; and two, to stimulate interest in the still wide-open field of topological transformation groups. The second more unexplored topic is the larger and has long been of interest to us.

The subject matter of this book originates in the nineteenth century concept of "finite continuous groups". These are families of many-times differentiable transformations of a region of the space of n real or complex numbers into another such region. These part-analytic, part-geometric structures are often only local groups (local Lie groups); the region of definition of the group operations is usually dependent on solutions of certain differential equations, so that there is actually given a "group-germ" rather than a group.

The present-day concept of a group (in the large) also possessing a topological structure developed around the end of the last century, roughly parallel to the development of the subject of topology. Of course in the work of Lie and others on "continuous groups" there appear many concepts and many types of arguments which we now call topological. The highlights of this development are perhaps these:

First, the background of work of Lie, and others.

Second, the formulation by Hilbert of his fifth problem (in his address at the International Congress of Mathematics, in 1900, setting forth twenty-three large programs of research) in which he underscored the class of locally euclidean groups as distinguished from groups with differentiability properties (see 2.15 of the present text); then also the work of Hilbert on the foundations of plane geometry from the point of view of topological properties of groups of congruences.

Third, perhaps, the work of Brouwer on transformation groups of the line and the plane, and his work on the cantor-discontinuum as a group manifold (1910—1912) coupled with his work on the topology of locally euclidean spaces. Then, through the work of E. Cartan and H. Weyl and others there resulted a very complete view of the Lie groups as geometric entities, and much information about their global structure, as well as their relation to subgroups of the matrix groups. What we may now call the classical period (preceding 1935) terminated with the analysis of compact groups by von Neumann, and of locally compact abelian groups by Pontrjagin, and a considerable extension of this work to other important classes of groups.

After 1935 the subject developed along many different and important lines, but we shall concentrate on only two of these: first, the relation of locally compact groups to Lie groups (extending the results already known for compact groups); and second, some selected topics in the theory of transformation groups which happen to be in the field of principal interest to the authors.

The first four chapters of the book deal with the first problem, and they are on the whole self-contained. We have not repeated the theory of differential equations underlying the introduction of analytic coordinates, nor the details of the theory of characteristic functions of integral equations, but we have given almost everything else that is needed in rather complete detail. The first chapter is a background course in the elementary topological notions which are constantly in use in the book. The second chapter reviews the classical period; this has been covered in the book of Pontrjagin and some of what we do is also in Weil, but is here repeated, with some new material.

Chapter III and part of IV deal with new methods by which the results of the earlier period have been extended to all locally compact groups. In many ways this work goes beyond the program suggested by Hilbert. However, many aspects of that program or its natural outgrowth remain open.

In the earlier chapters there is some material on transformation groups, and the last two chapters are largely given over to this concept. Here we have a topological group, usually a compact one, acting as a group of "motions" of some standard space, usually the n-space, n-sphere, or other n-manifold, and the guiding thought in the investigations may be said to be to find how nearly the situation resembles that of a group of linear transformations of the space or some other familiar group. It is only in the plane or in three-space that this problem is fairly completely solved.

The last two chapters rely on material not exposited in the book, using a few basic results from homology theory, dimension theory, and fibre mappings, as these are needed. Since this part of the field is not near its final form, it is perhaps better to leave to a later date an attempt at a complete and detailed exposition.

A part of this material was presented by one of us at the Summer Colloquium of the American Mathematical Society in Minneapolis in 1951, and was originally intended for publication in the Colloquium Series of the Society. This part has been completely revised and is being published here with new material.

The authors express their appreciation of the friendly care with which Professors R. Jacoby and E. Schenkman have read the manuscript, of the many valued suggestions they made, and of their help in the proofreading.

It is a pleasure to offer our thanks to W. Huebsch, M. Kuranishi, and C. T. Yang, who also helped with proofreading. One of the authors acknowledges also the invaluable assistance of the Office of Naval Research and National Science Foundation subsidies.

DEANE MONTGOMERY
LEO ZIPPIN

Institute for Advanced Study, Princeton, N. J.
The City University of New York, New York

For this reprinting we have corrected errors and misprints called to our attention by friends. We have also added a selected bibliography of some of the interesting recent developments in the field, many of which are concerned with differentiably acting groups.

CONTENTS

CHAPTER I

Topological Spaces and Groups

CHAPTER II

Locally Compact Groups

CHAPTER III

Groups with no Small Subgroups

CHAPTER IV

Approximation by Lie Groups

CHAPTER V

Transformation Groups

CHAPTER VI

Compact Transformation Groups

CHAPTER I

Topological Spaces and Groups

1.0. Introduction

This chapter contains the preliminary and somewhat elementary facts of general spaces and groups. Proofs are given in considerable detail and there are examples which may be of help to a reader for whom the subject is new.

We use the standard set-theoretic symbols: capitals A, B, etc. for *sets*, $A \cup B$ for the *union* of sets (elements in one or both), $A \cap B$ for the *intersection* (elements in both), etc.

1.1. Spaces

The term *space* is sometimes used in mathematical literature in a very general sense to denote any collection whose individual objects are called points, but in topology the term space is used only when some further structure is specified for the collection. As the term will be used in this book it has a meaning which is convenient in studying topological groups. The definition is as follows:

DEFINITION. *A topological space (or more simply space) is a non-empty set of points certain subsets of which are designated as open and where, moreover, these open sets are subject to the following conditions:*

1) *The intersection of any finite number of open sets is open.*
2) *The union of any number of open sets is open.*
3) *The empty set and the whole space are open.*
4) *To each pair of distinct points of space there is associated at least one open set which contains one of the points and does not contain the other.*

A space is called *discrete* if each point is an open set.

Condition 4) is known as the T_0-separation axiom in the terminology of Alexandroff and Hopf. The first three conditions define a topological space in their terminology. The designated system of open sets is the essential part of the topology, and the same set of points can become a topological space in many ways by choosing different systems of subsets designated as open.

1.2. Homeomorphisms

DEFINITION. *A homeomorphism is a one-one relation between all points of one topological space and all points of a second which puts the open sets of the two spaces in one-one correspondence; the spaces are topologically equivalent.*

The notion of homeomorphism is reflexive, symmetric, and transitive so that it is an equivalence relation in a given set of topological spaces.

Examples of spaces. Let E_1 denote the set of all real numbers in its customary topology: the open intervals are the sets $\{y; x < y < z\}$ for every $x < z$. The open sets are those which are unions of open intervals together with the empty-set (null-set) and the whole space.

Let $R_1 \subset E_1$ denote the set of numbers in the closed interval $0 \leqq y \leqq 1$, where for the moment we take this subset without a topology. This set gives distinct spaces as follows:

1) Topologize R_1 as customarily: the open sets are the intersections of R_1 with the open sets of E_1.

2) Topologize R_1 discretely, that is let every subset be open.

3) Topologize R_1 by the choice: open sets are the null set, the whole space and for each x of R_1 the set $\{z, x < z \leqq 1\}$.

4) Topologize R_1 by the choice: the complement of any finite set is open and the empty set and the whole space are open.

In the sequel R_1 will denote the closed unit interval and E_1 the set of all reals in the customary topology. A set homeomorphic to R_1 is called an *arc*. A set homeomorphic to a circle is called *a simple closed curve*.

1.3. Basis

DEFINITION. *A collection $\{Q_a\}$ of open sets of a space is called a basis for open sets if every open set (except possibly the null set) in the space can be represented as a union of sets in $\{Q_a\}$. It is called a sub-basis if every open set can be represented as a union of finite intersections of sets in $\{Q_a\}$ (except possibly the null set).*

A collection $\{Q_a\}$ of open sets of a space S is a basis if and only if for every open set Q in S and $x \in Q$ there is a $Q_a \in \{Q_a\}$ such that

$$x \in Q_a \subset Q.$$

If a collection has this property at a particular point x then the collection is called a *basis at x*.

If a set together with certain subsets are called a sub-basis, then another family of subsets is determined from the sub-basis by taking arbitrary unions and finite intersections. This new family (with the null set added if necessary) then satisfies conditions 1), 2), 3) for the open sets of a topological space. Whether 4) will also be satisfied depends on the original family of sets.

EXAMPLE. Let E_1 denote the space of real numbers in its usual topology. For each pair of rationals $r_1 < r_2$ let (r_1, r_2) denote the set of reals $r_1 < x < r_2$. This countable collection of open sets is a basis.

A space is said to be *separable* or to satisfy the *second countability axiom* if it has a countable basis. A space is said to satisfy the "first countability axiom" if it has a countable basis at each point.

EXAMPLE. Let S denote a topological space and let F denote a collection of real valued functions $f(x)$, $x \in S$. If f_0 is a particular element of F then for each positive integer n let $Q(f_0, n) = \{f \in F; \ |f(x) - f_0(x)| < 1/n$ for all $x\}$. We may topologize F by choosing the sets $Q(f_0, n)$ for all f_0 and n as a *sub-basis*. The topological space so obtained has a countable basis at each point in many important cases.

1.4. Topology of subsets

Let S be a topological space, T a subset. Let $Q \cap T$ be called *open in* T or open *relative* to T if Q is open in S. With open sets defined in this way T becomes a topological space and the topology so defined in T is called the *induced* or *relative* topology. If S has a countable basis and $T \subset S$ then T has a countable basis in the induced topology.

DEFINITION. *A subset $X \subset S$ is called closed if the complement $S - X$ is open. If $X \subset T \subset S$, X is called closed in T when $T - X$ is open in T.*

Notice that T closed in S and X closed in T implies that X is closed in S. The corresponding assertion for relatively open sets is also true.

It can be seen that finite unions and arbitrary intersections of closed sets are again closed.

DEFINITION. *If $K \subset S$, the intersection of all closed subsets of S which contain K is called the closure of K and is denoted by $\overline{K}$. If K is closed, $K = \overline{K}$.*

1.5. Continuous maps

Let S and T be spaces (= topological spaces) and f a map of S into T

$$f : S \to T;$$

that is, for each x in S, $y = f(x)$ is a point of T. If the inverse of each relatively open set in $f(S)$ is an open set in S then f is called *continuous*. In case $f(S) = T$ then f continuous and V open in T imply $f^{-1}(V)$ is open in S. The map is called an *open map* if it carries open sets to open sets.

If f is a continuous map of S *onto* T (that is $f(S) = T$) and if f^{-1} is also single valued and continuous, then f and f^{-1} are homeomorphisms and S and T are homeomorphic or topologically equivalent (1.2).

EXAMPLE. The map $f(t) = \exp(2\pi\sqrt{-1}\,t)$ is a continuous and open map of E_1 onto a circle (circumference) in the complex plane.

EXAMPLE. Let K denote the cylindrical surface, described in x, y, z coordinates in three-space by $x^2 + y^2 = 1$. Let f_1 denote the map of K onto E_1 given by $(x, y, z) \rightarrow (0, 0, z)$, let f_2 denote the map of K that is given by $(x, y, z) \rightarrow (x, y, 0)$ and f_3 the map $(x, y, z) \rightarrow (x, y, |z|)$ of K into K. All three maps are continuous, the first two are open, and f_1 and f_3 are also *closed*, i.e. they map closed sets into closed sets.

1.6. *Topological products*

The space of n real variables $(x_1, x_2, \ldots, x_n)$, $-\infty < x_i < \infty$, $i = 1, \ldots, n$, and the cylinder K of the preceding example are instances of *topological products*.

Let A denote any non-null set of indices and suppose that to each $a \epsilon A$ there is associated a topological space S_a. The totality of functions f defined on A such that $f(a) \epsilon S_a$, for each $a \epsilon A$, is called the *product* of the spaces S_a. When topologized as below it will be denoted by PROD S_a; we also use the standard symbol $\times$, thus $E \times B$ is the set of ordered pairs (e, b), $e \epsilon E$, $b \epsilon B$.

The standard topology for this product space is defined as follows. For each positive integer n, for each choice of n indices $a_1, a_2, \ldots, a_n$, and for each choice of a non-empty open set in S_{a_i}

$$U_{a_i} \subset S_{a_i}, \ i = 1, 2, \ldots, n,$$

consider the set of functions $f \epsilon$ PROD S_a for which

$$f(a_i) \epsilon U_{a_i}, \ i = 1, \ldots, n.$$

Let the totality of these sets be a sub-basis for the product. The resulting family of open sets satisfies the definition of space in 1.1.

EXAMPLE 1. The space $E_n = E_1 \times E_1 \times \ldots \times E_1$, n copies, is the space of n real variables; here $A = \{1, 2, \ldots, n\}$ and each S_i is homeomorphic to E_1 (1.2). Let $x_i \epsilon S_i$. Then $(x_1, \ldots, x_n)$

are the "coordinates" of a point of E_n. It can be verified that the sets $U_m(x)$, $m \in I$ (the collection of positive integers), of points of E_n whose euclidean distance from $x = (x_1, \ldots, x_n)$ is less than $1/m$, form a basis at x. The subset $R_1 \times R_1 \times \ldots \times R_1$ is an n-cell.

EXAMPLE 2. Let A be of arbitrary cardinal power and let each S_a, $a \in A$, be homeomorphic to C_1, the circumference of a circle. Then PROD S_a is a *generalized torus*. If A consists of n objects, the product-space is the *n-dimensional* torus. For $n = 2$, we get the *torus*.

EXAMPLE 3. Let $D = S_1 \times S_2 \times \ldots \times S_n \times \ldots$, $n \in I$, where each S_i is a pair of points — conveniently regarded as the "same" pair, and designated 0 and 2. This is the Cantor Discontinuum, or Cantor Middle Third Set. It is homeomorphic to the subset of the unit interval defined by the convergent series: $D : \{\Sigma a_n/3^n\}$, $a_n = 0$ or 2. This example will be described in another way in the next section.

THEOREM: *Let F_a be a closed subset of the topological space S_a, $a \in A$. Then PROD F_a is a closed subset of PROD S_a.* The proof is left to the reader.

1.7. Compactness

DEFINITION. *A topological space S is compact if every collection of open sets whose union covers S contains a finite subcollection whose union covers S.*

EXAMPLE 1. The unit interval R_1 is compact. Thus let $\{Q\}$ denote a collection of open sets covering R_1. Let F denote the set of points $x \in R$ such that the interval $0 \leq y \leq x$ can be covered by a finite subcollection of $\{Q\}$. Then F is not empty and is both open and closed. Hence by the Dedekind cut postulate, or the existence of least upper bounds, or the connectedness of R_1, it follows that $F = R_1$. To illustrate the concept of compactness consider the open sets $W_n \subset R_1$, $W_n : 1/3n < x < 1/n$,

$n \epsilon I$. This collection does not cover R_1. Let W_a be the union of two sets: $0 \leq x < a$ and $1 - a < x \leq 1$, for some $a, 0 < a < 1$. Now, no matter how $a > 0$ is chosen, there is always some finite number of the W_n which *together* with W_a covers R_1. Of course R_1 minus endpoints is not compact and no finite subcollection of the W_n in this example will cover it.

THEOREM. *Let S be a compact space and let $f : S \to T$ be a continuous map of S onto a topological space T. Then T is compact.*

Let $\{O_a\}$ be a covering of T by open sets. Since f is continuous, each $f^{-1}(O_a)$ is an open set in S. There is a finite covering of S by sets of the collection $\{f^{-1}(O_a)\}$, and this gives a corresponding finite covering of T by sets of $\{O_a\}$. This completes the proof.

COROLLARY. *If f is a continuous map of S into T then $f(S)$ is a compact subset of T.*

1.7.1. THEOREM. *Let S be a compact space and $\{D_a\}$ a collection of closed subsets such that $\cap_a D_a$ is empty. Then there is some finite set $D_{a_1}, \ldots, D_{a_n}$ such that $\cap_i D_{a_i}$ is empty.*

The complement of $\cap_a D_a$ is $\cup_a (S - D_a)$; if the intersection-set is empty, the union covers S. There is a finite set of indices a_i such that $S \subset \cup_i (S - D_{a_i})$ and consequently $\cap_i D_{a_i}$ is empty, for the same finite set of indices.

COROLLARY 1. *Let D_n, $n \epsilon I$, be a sequence of non-empty closed subsets of the compact space S, with $D_{n+1} \subset D_n$. Then $\cap_n D_n$ is not empty.*

APPLICATION: *The Cantor Middle Third Set D.* From R_1, "delete" the middle third: $1/3 < x < 2/3$. Let D_1 denote the residue: it is a union of two closed intervals. Let D_2 denote the closed set in D_1 complementary to the union of the middle third intervals: $1/9 < x < 2/9$ and $7/9 < x < 8/9$. Continuing inductively, define $D_n \subset D_{n-1}$ consisting of 2^n closed mutually exclusive intervals. Let $D = \cap D_n$. This is homeomorphic to the space of Example 3 of 1.6.

COROLLARY 2. *A lower semi-continuous (upper semi-continuous) real-valued function on a compact space has finite g. l. b., greatest lower bound (and l. u. b., least upper bound), and always attains these bounds at some points of space.*

This follows from the preceding corollary and the fact that the set where $f(x) \leqq r$ is closed, for every r (similarly, $f(x) \geqq r$).

1.7.2.   THEOREM.   *A topological space with the property*: "*every collection of closed subsets with empty set-intersection has a finite subcollection whose set-intersection is empty*", *is compact.*

The proof, like that of the Theorem of 1.7.1, is based on the duality between open and closed sets.

DEFINITION.   *If a point x of a topological space S belongs to an open subset of S whose closure is compact, then S is called locally compact at x; S is locally compact if it has this property at every point.*

COROLLARY.   *A closed subset of a locally compact space is locally compact in the induced topology.   Similarly a closed subset of a compact space is compact.   The union of a finite number of compact subsets is compact.*

The proof is left to the reader.

A set U in a topological space is called a *neighborhood* of a point z if there is an open set O such that $z \in O \subset U$; z is called an *inner point* of U.   A set F is covered by a collection $\{U_i\}$ if each point of F is an inner point of some set U_i.

1.7.3.   A space S is called a *Hausdorff space* if for every $x, y \in S$, $x \neq y$, there exist open sets U and V including x and y respectively such that $U \cap V = \Phi$ where Φ is the empty-set; an equivalent property is the existence of a closed neighborhood of x not meeting y.   It is left to the reader to show that *a compact subset of a Hausdorff space is closed.*

LEMMA.   *Let S be a compact Hausdorff space, let F be a closed set in S, and x a point not in F.   Then there is a closed neighborhood W of x such that $W \cap F = \Phi$.*

For each $y \in F$ let U_y be a neighborhood of y and W_y a neighborhood of x, such that $U_y \cap W_y = \Phi$.   There is a covering of F

by sets $U_{v_1}, \ldots, U_{v_n}$. Let W_x be the intersection of the associated W_{v_i}, $i = 1, \ldots, n$, and let W be the closure of W_x. The union of the U_{v_i} does not meet W. Then $W \cap F = \Phi$, which completes the proof.

A space with a property of this kind is called *regular*. A regular space in which each point is a closed set is also a Hausdorff space.

THEOREM. *Let U be a compact Hausdorff space and let F_n be a sequence of closed subsets of U. If U is contained in the union of sets F_n then at least one of the sets F_n has inner points.*

Take a sequence $C_1 \supset C_2 \supset \ldots$, of non-empty compact neighborhoods such that for each n, $(\cup_1^n F_i) \cap C_n = \Phi$. This leads to $\cap C_n = \Phi$, a contradiction.

A set is *nowhere dense* if its closure has no inner points. A space is said to be of the *second category* if it cannot be expressed as the union of a countable number of nowhere dense subsets. Hence a *compact Hausdorff space* is of the *second category*. Complete metric spaces, to be defined later, are also of the second category.

1.7.4. THEOREM. *Let S be a locally compact space. There exists a compact space S^* and a point z in S^* such that $S^* - z$ is homeomorphic to S.*

Let z denote a "new" point, not in S, and let S^* denote the set-union of S and z. If Y is a subset of S, let $Y^* \subset S^*$ denote the union of Y and z. We topologize S^* as follows. Any open set in S is also open in S^*. In addition, if X is a compact subset of S and Y is the complement $S - X$, then Y^* is open in S^*. These open sets are taken as a sub-basis for open sets in S^*.

Suppose now that we have some covering of S^* by a family of open sets. Then z belongs to one of these open sets, say $z \, \varepsilon \, U^*$. The complement of U^* is a compact subset of S. Hence the complement is covered by a finite subset of the given covering sets, because of the compactness in S. Together with U^*, this gives a finite covering of S^*.

1.8. Tychonoff theorem

THEOREM. *Let S_a, $a \, \varepsilon \, \{a\}$, be compact spaces and let P be the topological product of the S_a. Then P is compact.*

We shall first prove this theorem for the case of *two* factors. In the next section we shall prove the general case without reference to the number of factors.

Let $P = S_1 \times S_2$, and let F denote a family of open sets of P covering P. For each point x_1 of S_1, the closed subset $x_1 \times S_2$ of P is homeomorphic to S_2 and is therefore compact. Each point of $x_1 \times S_2$ belongs to a set in F because F is a covering. Because of the way in which a product is topologized, it follows that each point of $x_1 \times S_2$ belongs to some open set $U \times V$ of P such that $U \times V$ is a subset of some set of F. It follows from its compactness that $x_1 \times S_2$ is contained in the union of a finite number of sets

$$U_1 \times V_1, \ldots, U_n \times V_n,$$

each of which is a subset of some set of F. Let $U' = \cap U_i$. Then $U' \times S_2$ is covered by a finite number of sets of F.

Since x_1 is an arbitrary point of S_1 and S_1 is compact, there exists a finite number of open sets of S_1

$$U_1', \ldots, U_m'$$

which cover S_1 and which are such that there is a finite number of sets of F covering $U_i' \times S_2$, $i = 1, \ldots, m$. The totality of sets of F thus indicated is a finite number which covers $S_1 \times S_2$. This completes the proof for the case of two factors. The case for a finite number of factors follows by a simple induction.

EXAMPLE. Let $R_n = R_1 \times R_1 \times \ldots \times R_1$, n factors. Then R_n is compact and it follows that E_n ($=$ product of n real lines) is locally compact.

1.8.1. To consider the general case, let $\{a\}$ be an arbitrary collection of at least two indices: let S_a be compact topological spaces, let P be the topological product, and let F be a collection of open subsets of P covering P. The proof that P is compact is by contradiction. Accordingly, we shall suppose that *no finite subcollection of sets of F covers P*.

It was shown by Zermelo that it is possible to well-order the set of all subsets of a given set by the use of an axiom-of-choice of appropriate power, namely the cardinal number of the set of all subsets of the given set. A well-ordering of objects permits them to be inspected systematically.

Using such a well-ordering we can enlarge the given family F to a family F^* of open sets, where F^* has the following properties:

1) F^* is a covering of P by open sets.

2) No finite subcollection of F^* covers P.

3) If we adjoin to F^* any open subset of P not already in F^*, then the enlarged collection does contain a finite subcollection which covers P. Of course it is in 3) that F^* has a property not necessarily true of F.

Using this enlarged family the proof for the general case becomes similar to the proof for two factors. Let b denote an arbitrary index in $\{a\}$, which shall be fixed temporarily, and let P_b denote the product of all factors S_a excepting S_b. Then P is (homeomorphic to) $S_b \times P_b$.

Suppose for a moment that to each point $x_b \in S_b$ there exists an open $U_b \subset S_b$ containing x_b such that

$$*) \quad U_b \times P_b \in F^*.$$

There must then exist some finite covering of S_b by sets $U_b^1, U_b^2, \ldots, U_b^n$ each satisfying *). The product P is covered by the union of $U_b^i \times P_b$, $i = 1, \ldots, n$. This is impossible by the construction of F^*. Hence in each S_b there is at least one x_b which does not satisfy the first sentence of this paragraph.

It follows by the axiom of choice that P contains at least one point $x = \text{PROD } x_b$ such that if U_b is an open set in S_b and $x_b \in U_b$ then *) is false. This holds for each coordinate x_b of x. This implies for each coordinate x_b of x that if x_b is in an open set U_b of S_b then there is *a finite collection of sets in F^**

$$O_b^1, O_b^2, \ldots, O_b^{n_b}$$

which together with $U_b \times P_b$ forms a covering of P.

The point x belongs to an open set $O_x \in F^*$. There is some open set contained in O_x which contains x and is of the form

$$U_{a_1} \times U_{a_2} \times \ldots \times U_{a_n} \times P_{a_1 a_2 \ldots a_n}$$

for some finite set of indices a_i, and where the last set is the product of all S_a with the exception of S_{a_i}, $i = 1, 2, \ldots, n$. For each a_i there exists a finite collection of sets of F^* which together with $U_{a_i} \times P_{a_i}$ covers P, say these sets are

**) $\quad O_{a_i}^1,\ O_{a_i}^2,\ \ldots,\ O_{a_i}^{n_i},\ i = 1, 2, \ldots, n.$

Then P is covered by the union of O_x and the sets of **). This contradiction completes the proof.

1.8.2. EXAMPLE. The infinite-dimensional torus described in 1.6 whose "dimension" equals the cardinal power of the set of indices A is compact. It is a commutative group where the addition of two points is carried out by adding the respective coordinates in each factor $S_a = C_1$, each of these factors being itself a commutative group. The group addition is continuous in the topology and this defines a topological group (1.11). In fact this is a universal compact commutative topological group (depending on the cardinal power of the group). See Alexander-Zippin [1].

The principal theorem of this section is due to Tychonoff [1]. The present proof is dual to a proof given by Bourbaki [1].

1.9. Metric spaces

DEFINITION. A set S of points is called a *metric space* if to each pair $x, y \in S$ there is associated a non-negative real number $d(x, y)$, the *distance* from x to y, satisfying

    1) $\quad d(x, y) = 0$ if and only if $x = y$

    2) $\quad d(x, y) = d(y, x)$

    3) $\quad d(x, y) + d(y, z) \geq d(x, z),\ x, y, z \in S.$

The *distance function* $d(x, y)$, also called the *metric*, induces a topology in S as follows. For each $r > 0$ let $S_r(x)$ denote the *sphere* of radius r, i.e. the set of $y \in S$ such that $d(x, y) < r$. Now let $S_r(x)$, for all positive r and all $x \in S$ constitute a basis for open sets. This choice of basis makes S a topological space. A space is called *metrizable* if a metric can be defined for it which induces in it the original topology. It is clear that a metric space has a countable basis at each point x, namely $S_r(x)$, r rational.

EXAMPLE 1. If S_1 and S_2 are metric spaces then $S_1 \times S_2$ is a metric space in the metric

$$d((x_1, x_2), (y_1, y_2)) = \text{Max } (d_1(x_1, y_1), d_2(x_2, y_2))$$

where x_1, $y_1 \in S_1$, x_2, $y_2 \in S_2$. The topology determined by this metric is the same as the product topology.

EXAMPLE 2. The set F of continuous functions defined on a compact space S with values in a metric space M becomes a metric space by defining for f, g in F

$$d(f, g) = \text{l.u.b. } (x \in S) [d_M(f(x), g(x))]$$

where d_M is the metric in M. (See corollary 2, section 1.7.1.)

THEOREM. *The collection of open sets of a compact metric space S has a countable basis.*

For each $n \in I$ there is a covering of the space by a finite number of open sets each of diameter at most $1/n$. The countable collection of these sets for all n is a basis.

EXAMPLE 3. If S is a compact metric space and E_1 denotes the real line, then $S \times E_1$ is a metrizable, locally compact space with a countable basis for open sets.

By Example 1 above the space is metrizable. If $\{U_m\}$ and $\{V_n\}$ are countable bases in S and E_1 respectively then $\{U_m \times V_n\}$ forms a countable base in $S \times E_1$. If $E_{1n} = \{x \in E_1, |x| \leq n\}$ then $S \times E_{1n}$ is a compact subset of $S \times E_1$ and any point of the product is interior to $S \times E_{1n}$ for n large enough. This proves the local compactness.

1.9.1. The following is of interest: *If (x, y) is a metric for a space M then the following equivalent metric:*

$$(x, y)' = \frac{(x, y)}{1 + (x, y)} \leq 1$$

is a bounded metric. Properties 1) and 2), above, are obviously satisfied. For 3), one uses the fact that the function $t/(1 + t)$ increases with t. Thus:

$$(x, y)' + (y, z)' \geqq \frac{(x, y)}{1 + (x, y) + (y, z)} + \frac{(y, z)}{1 + (y, z) + (x, y)}$$

$$= \frac{(x, y) + (y, z)}{1 + (x, y) + (y, z)} \geqq \frac{(x, z)}{1 + (x, z)} = (x, z)'.$$

This has the following consequence.

LEMMA. *Let M be a space which is the union of a system M_a $a \in \{a\}$, of open mutually exclusive sets. Suppose each M_a is a metric space and carries a metric d_a bounded by 1. Define a function $d(x, y)$ which is equal to 2 if x and y are not in the same M_a; otherwise let d agree with the appropriate d_a. Then d is a metric for M.*
We omit the easy proof.

1.9.2.   A sequence of points x_n in a metric space is said to *converge* to a point x, symbolically $x_n \to x$, if $\lim d(x, x_n) = 0$. A sequence of points x_n satisfies the *Cauchy convergence criterion* if when $\epsilon > 0$ is given there is an N such that for $m, n > N$, $d(x_n, x_m) < \epsilon$. A metric space is called *complete* if every sequence of points satisfying the Cauchy condition converges to a point of the space.   A subset of a space is called *dense* (*everywhere-dense*) in the space, if every point of space is a limit of some sequence of points of the subset.

1.10. Sequential convergence

The proof of the principal theorem of this section illustrates a standard technique.   It will involve choosing an infinite sequence, then an infinite subsequence, then again an infinite subsequence and so on repeating this construction a countably infinite number of times.   A special form of this method is called the *Cantor diagonalizing procedure*. To facilitate the working of this technique we shall sometimes use the following notation.

1.10.1.   The letter I will denote the sequence of natural numbers $1, 2, 3, \ldots$.   When subsequences of I need to be chosen they will be labeled in some systematic way: $I_1, I_2, \ldots$ or $I', I'', \ldots$ or $I^*, I^{**}, \ldots$ and so on. Then given a sequence of elements: x_n, $n \in I$, we can refer to a subsequence as: x_n, $n \in I_1$, or: x_n, $n \in I^*$ and so on.

1.10.2. If A is a subset of a metric space M let $S_\epsilon(A) = \cup_x S_\epsilon(x)$, $x \in A$, where as above $S_\epsilon(x) = \{y \in M, \ d(x, y) < \epsilon\}$.

DEFINITION. Let S denote a metric space and let K_n, $n \in I$, be a sequence of subsets of S. The sequence K_n is said to *converge* to a set K if for every $\epsilon > 0$

1.10.3. $K_n \subset S_\epsilon(K)$ *and* $K \subset S_\epsilon(K_n)$

for n sufficiently large (depending only on ϵ).

If K_n is given and if a K exists satisfying relation 1.10.3 then $\overline{K}$ also satisfies 1.10.3. If two closed sets K' and K'' satisfy 1.10.3 for the same sequence K_n then $K' = K''$. In the special case that the sets K_n are single points, the set K if it exists is a point and is unique.

1.10.4. THEOREM. *Every sequence of non-empty subsets of a compact metric space S has a convergent subsequence.*

Let K_n, $n \in I$, be an arbitrary sequence of subsets of S and let W_n, $n \in I$, be a basis for open sets in S (Theorem 1.9).

Let I be called I_0 and suppose a sequence I_{m-1} has been defined. Consider $W_m \cap K_n$, $n \in I_{m-1}$, m fixed. Then either $W_m \cap K_n$ is not empty for an infinite subsequence of integers $n \in I_{m-1}$ or on the contrary $W_m \cap K_n$ is empty for almost all $n \in I_{m-1}$. In the first of these cases define I_m as the set of indices n in I_{m-1} such that $W_m \cap K_n$ is not empty; and in the second case define I_m so that $W_m \cap K_n$ is *always* empty for $n \in I_m$. Then in all possible cases $I_m \subset I_{m-1}$ is uniquely defined. We now consider I_m to be defined by induction for all $m \in I$.

We can now specify what subsequence of K_n we may take as convergent. Let $I^* \subset I$ denote the *diagonal sequence* of the sequences I_m, that is I^* contains the m-th element of I_m for each m. We shall show that K_n, $n \in I^*$, is convergent. It follows from the definition of $I*$ that for each $m \in I$, if we except at most the first m integers in $I*$,

$$W_m \cap K_n, \quad n \in I*$$

is always empty or is never empty depending on m.

We next define the set K to which the sets K_n, $n \in I*$, will be shown to converge. Let W denote the union of those sets W_m, $m \in I$ for which $W_m \cap K_n$, $n \in I*$, is almost always empty. Let $K = S - W$. Since each W_m forming W meets at most a finite number of the sets K_n, no finite number of these W_m can cover S. Therefore W cannot cover S, since S is compact, and hence K is not empty. The set K is closed and therefore compact.

Let $\epsilon > 0$ be given. There is a covering of K by sets

$$W_{k_1}, W_{k_2}, \ldots, W_{k_s}, \quad k_i \in I$$

each meeting K and each of diameter less than ϵ. None of the sets W_{k_i} can belong to W and each must intersect almost all the K_n, $n \in I*$. Therefore for sufficiently large n, $n \in I*$

$$W_{k_i} \cap K_n \neq \Phi.$$

It follows that

$$K \subset S_\epsilon(K_n), \quad n \in I*,$$

for all n sufficiently large. Finally it can be seen that the closed set $S - S_\epsilon(K)$ is contained in W. It follows that the complement of $S_\epsilon(K)$ is covered by sets

$$W_{j_1}, W_{j_2}, \ldots, W_{j_t}$$

each of which is an element in the union defining W. Therefore

$$W_{j_k} \cap K_n, \quad n \in I*, \quad k = 1, 2, \ldots, t$$

is almost always empty. Therefore for sufficiently large $n \in I*$

$$K_n \subset S_\epsilon(K).$$

This completes the proof of the Theorem.

1.10.5.   Let X and Y be closed subsets of a compact metric space S. Define the *Hausdorff metric*: $d(X, Y)$ as the greatest lower bound of all ϵ such that symmetrically

$$X \subset S_\epsilon(Y), \quad Y \subset S_\epsilon(X).$$

This is a metric for the collection of closed subsets of S.

THEOREM. *The set F of all closed subsets of a compact metric space S is a compact metric space in the metric defined above.*

The set F is a metric space in the metric defined above. If A_n is in F then A_n has a subsequence converging to a set A in the sense of convergence defined above and the set A may be assumed closed. It follows that the subsequence also converges to A in the sense of the metric of F. Hence every sequence in F has a convergent subsequence.

Let W_n, $n \in I$, be a basis for open sets in S as in the preceding Theorem. For each n and each choice of integers $k_1, \ldots, k_n$, let $W(k_1, \ldots, k_n)$ denote the union of the sets $W_{k_1}, \ldots, W_{k_n}$. Now in F let $W^*(k_1, \ldots, k_n)$ consist of all the compact sets in S which belong to $W(k_1, \ldots, k_n)$ and meet each W_{k_i}. This gives a countable collection of subsets of F. The proof of the preceding theorem shows that this collection is a basis for open sets in F.

Finally, let $\{O_n\}$, $n = 1, 2, \ldots$, be a *countable* collection of open sets of F which cover F. We have to find an integer m such that $\bigcup_1^m O_i$ covers F. If no such integer existed we could find a sequence of points x_n, $x_n \varepsilon F - \bigcup_1^n O_i$. This sequence would have to have a subsequence converging to some point x. Since $x \in O_m$ for some m, it follows that infinitely many of the x_n belong to O_m; this contradiction proves the theorem.

1.10.6. Note that separability implies that every collection of covering sets has a countable covering subcollection. It also implies that there exists a countable set of points which is everywhere dense in the space.

THEOREM. *A metric space S is compact if and only if every infinite sequence of points has a convergent subsequence.*

Suppose that every infinite subsequence of points of S has a convergent subsequence. We shall prove that S is separable.

The last paragraph of the preceding section then shows that S is compact. The converse is shown in 1.10.4.

For each positive integer n construct a set P_n such that 1) every point of P_n is at a distance at least $1/n$ from every other point of P_n, 2) every point of S not in P_n is at a distance less than $1/n$ from some point of P_n. It is easy to see that no sequence of points in any one P_n can be convergent, and it follows that P_n is a *finite* point set. Let $P = \cup P_n$. Then P is countable, and every point of S is a limit point of P. For each rational $r > 0$ and each point of P construct the "sphere" with that point as center and radius r. The set of these spheres is countable, and is a basis for open sets. This concludes the proof.

EXAMPLE. Let S be a compact metric space let H be the space of real continuous functions defined on S, with values in R_1. Each continuous function $f(x)$ determines a closed subset of $S \times R_1$, namely the graph consisting of the pairs $(x, f(x))$, $x \varepsilon S$. Hence H is a *subset* of a compact metric space (see examples 2 and 3 in 1.9). Notice that the metric defined for H in example 2 of 1.9 and the metric which it gets from $S \times R_1$ are topologically equivalent. This shows that H itself is a separable metric space.

1.11. Topological groups

Topological groups were first considered by Lie, who was concerned with groups defined by analytic relations (see 2.0). Around 1900—1910 Hilbert [1.2] and Brouwer [1.3] and others were interested in more general topological groups. Brouwer [3] showed that the Cantor middle third set can be made into an abelian topological group. Later Schreier [1] and Leja [1] gave a definition in terms of topological spaces whose theory had been developed in the intervening time.

A topological group is a topological space whose points are elements of an abstract group, the operations of the group being continuous in the topology of the space. A detailed definition containing some redundancies is as follows:

DEFINITION. *A topological group G is a space in which for x, $y \in G$ there is a unique product xy in G and*

1) *There is a unique identity element e in G such that $xe = ex = x$ for all $x \in G$.*

2) *To each $x \in G$ there is an inverse x^{-1} such that $xx^{-1} = x^{-1}x = e$.*

3) *$x(yz) = (xy)z$ for x, y, $z \in G$.*

4) *The function x^{-1} is continuous on G and xy is continuous on $G \times G$.*

Familiar examples are the real or complex numbers under addition with the usual topologies for E_1 and E_2 respectively, and the complex numbers of absolute value one under multiplication with their usual topology as a subset of E_2. The space of this last group is homeomorphic to the circumference of a circle.

1.11.1. Of course, properties 1), 2), 3) define a group in the customary sense and a topological group may be thought of as a set of elements which is both an abstract group and a space, the two concepts being united through 4). When a subset H of G is itself a group then we shall call H a *subgroup* of G but we shall understand that H is to be given the relative topology. It is easy to see that then H becomes a topological group.

If H is a subgroup of G and if x and y are points of G belonging to the closure of H, then every neighborhood of the product element xy contains points of H. *For* let U be a neighborhood of xy. Then by 4) there exist neighborhoods V of x and W of y such that every product of an element of V and an element of W is contained in U. We see from this that xy belongs to the closure of H. Similarly, x^{-1} belongs to the closure. Thus, $\overline{H}$ is a group, and we shall call it a *closed* subgroup.

1.11.2. If G_a, $a \varepsilon \{a\}$, is a collection of topological groups and if G denotes the product space defined in 1.6, then G can be regarded as a group (the product of two elements of G being defined by the product of their components in each factor G_a). Because each neighborhood of the PROD G_a depends on only a finite number of the factors, it is easy to see that G becomes a topological group. We shall call it the topological product group of the factors G_a.

By way of examples, note that the product of an arbitrary number of groups each isomorphic to C_1: the group of reals-modulo-one (isomorphic to the complex numbers of modulus one under multiplication) is a compact topological group. The product of an arbitrary number of factors each isomorphic to the group of reals is a topological group which is locally compact if the number of factors is finite.

A finite group with the discrete topology is compact and the topological product of any collection of finite groups is therefore compact.

1.11.3. If x and y are in a topological group, $x \neq y$, then as will be seen (1.16) we may choose a neighborhood W of e such that

$$y \notin WW^{-1}.$$

Hence yW and xW are disjoined, and thus two distinct points of a topological group are in disjoined open sets. This is called the Hausdorff property (see 1.16); it implies that a point is a closed set.

It is easy to see that if H is an abelian subgroup of a topological group G then $\bar{H}$ is also abelian. Thus if x, $y \varepsilon \bar{H}$ and $xy \neq yx$ there are neighborhoods U_1 of xy and U_2 of yx with $U_1 \cap U_2 = \Phi$. There exist neighborhoods V of x and W of y such that for every $v \varepsilon V$ and $w \varepsilon W$, $vw \varepsilon U_1$ and $wv \varepsilon U_2$. However, if v, $w \varepsilon H$ then $wv = vw$ and we are led to a contradiction, proving that the closure of H is abelian.

1.11.4. Important examples of topological groups are given below.

EXAMPLE 1. The sets $M_n(R)$ and $M_n(C)$ of all $n \times n$ matrices of real and complex elements under addition with the distance of $A = (a_{ij})$, $B = (b_{ij})$ defined by

$$d(A, B) = \max_{i, j} |a_{ij} - b_{ij}|.$$

The spaces of these two groups are homeomorphic to E_{n^2} and E_{2n^2}. They are in fact the sets of real or complex vectors with

n^2 coordinates, and hence are vector spaces as well as groups. Another example is the set H of continuous real valued functions on a compact metric space under addition.

EXAMPLE 2. The sets of non-singular real or complex $n \times n$ matrices $Gl(n, R)$, $Gl(n, C)$ under multiplication; these are subsets of $M_n(R)$ and $M_n(C)$ respectively and are given the induced topology. They are open subsets and are therefore locally compact and locally euclidean (1.27).

EXAMPLE 3. Let S be a compact metric space and let G be the group of all homeomorphisms of S onto itself topologized as a subspace of the space of continuous maps of S into itself (1.9 example 2).

1.12. Isomorphism of topological groups

The spaces associated with two topological groups may be homeomorphic but the groups essentially different, for example, one abelian (= commutative) and the other not.

EXAMPLE 1. The matrices

$$\begin{pmatrix} a & 0 \\ 0 & b \end{pmatrix}, \qquad a, b \text{ real}$$

under addition. This is an abelian group with E_2 as space.

EXAMPLE 2. The matrices

$$\begin{pmatrix} e^a & b \\ 0 & e^{-a} \end{pmatrix} \qquad a, b \quad \text{real}$$

under multiplication. This is a non-abelian group with E_2 as space.

If we give the space in this example (or example 1) the discrete topology we obtain a new topological group with the same algebraic structure.

EXAMPLE 3. In the additive group of integers, for each pair of integers h and k, $k \neq 0$, let the set $\{h \pm nk\}$, $n = 0, 1, 2, \ldots$, be called an open set and let the collection of all these sets be taken as a basis for open sets.

EXAMPLE 4. Introduce a metric into the additive group of integers, depending on the prime number p, defined thus:

$$d(a, b) = 1/p^n$$

if $a \neq b$ and p^n is the highest power of p which is a factor of $a - b$.

EXAMPLE 5. Let G be the integers under addition with any set called open if it is the complement of a finite set (or is the whole space or the null set). Algebraically G is a group and it is also a space. However it is not a topological group because addition is now not simultaneously continuous. It is true however that addition is continuous in each variable separately.

For some types of group spaces separate continuity implies simultaneous continuity. It is not known whether this is true for a compact Hausdorff group space.

DEFINITION. *Two topological groups will be called isomorphic if there is a one-one correspondence between their elements which is a group isomorphism (preserves products and inverses) and a space homeomorphism (preserves open sets).*

An isomorphic map of G onto G is called an *automorphism*.

In examples 3) and 4) the abstract group structure of the additive group of integers is embodied in infinitely many non-isomorphic topological groups.

1.13. Set products

If G is a group, $A \subset G$, let A^{-1} denote the *inverse set*: $\{a^{-1}\}$ $a \in A$. Clearly $(A^{-1})^{-1} = A$. If $B \subset G$ let AB denote the set $\{ab\}$ $a \in A$, $b \in B$. It is understood that the *product set* is empty if either factor is empty. We shall write $AA = A^2$ and so on. It can be seen that $(AB)C = A(BC)$, and $(AB)^{-1} = B^{-1}A^{-1}$. The set AA^{-1} satisfies

$$(AA^{-1})^{-1} = AA^{-1}$$

that is, it is *symmetric*. Similarly the set intersection $A \cap A^{-1}$ is symmetric. The intersection of symmetric sets is symmetric.

A set H in G is called *invariant* if $gH = Hg$ for every $g \in G$, equivalently if $gHg^{-1} = H$.

THEOREM 1. *Let G be a topological group and let $A \subset G$ be an open set. Then A^{-1} is open.*

Let a^{-1} be in A. By the continuity of the inverse there exists an open set B containing a such that $b \in B$ implies $b^{-1} \in A$. This means that $B^{-1} \subset A$ and therefore $B \subset A^{-1}$. Thus A^{-1} is a union of open sets and is open.

COROLLARY. *The map $x \to x^{-1}$ is a homeomorphism.*

LEMMA. *Let G be a topological group, A an open subset, b an element. Then Ab and bA are open.*

Let $a \in A$ and let $c = ab$. Then $a = cb^{-1}$. Because a regarded as a product is continuous in c there must exist an open set C containing c such that if $c' \in C$ then $c'b^{-1} \in A$. But then $c' \in Ab$ and it follows that $C \subset Ab$. Therefore Ab is a union of open sets and is open. The proof that bA is open is similar.

COROLLARY. *For each $a \in G$ the left and right translations: $x \to ax$, $x \to xa$ are homeomorphisms.*

THEOREM 2. *Let G be a topological group and let A and B be subsets. If A or B is open then AB is open.*

Since AB is a union of sets of the form Ab, $b \in B$ it is open if A is open. Similarly AB is a union of sets aB, $a \in A$, and is open if B is open.

COROLLARY. *Let A be a closed subset of a topological group. Then Ab and bA are closed.*

This is true because left and right translations by the constant b are homeomorphisms of G onto G.

A function $f(x)$ taking a group G_1 into another group G_2 will be called a *homomorphism* if

$$*) \qquad f(x) f(y) = f(xy) \qquad x, y \in G_1.$$

When G_1 and G_2 are topological we shall ordinarily require that f be continuous. The most useful case is where f is open as well as continuous. In many situations (see 1.26.4 and 2.13) continuity implies openness but this is not true in general. The set of elements going into e is a subgroup and if f is continuous it is a closed subgroup (a point in a topological group is a closed set). This is the *kernel* of the homomorphism.

EXAMPLE: Let V_1 be the additive group of real numbers and G a topological group. A continuous homomorphism, $h(t)$, of V_1 into G, is called a *one-parameter group* in G. If h is defined only on an open interval around zero satisfying *) so far as it has meaning, then $h(t)$ is called a local one-parameter group in G. If $h(t)$ is a one-parameter group, the image of V_1 may consist of e alone and then $h(t)$ is a trivial one-parameter group. If this is not the case and if for some $t_1 \neq 0$, $h(t_1) = e$, then the image of V_1 is homeomorphic to a circumference. In case $h(t) = e$ only for $t = 0$, the image of V_1 is a one-one image of the line which may be a homeomorphism of the line or a very complicated imbedding of the line. To illustrate this let G be a torus which we obtain from the plane vector group V_2 by reducing mod one in both the x and y directions. In V_2 any line through the origin is a subgroup isomorphic to V_1 and after reduction the line $y = ax$ is mapped onto the torus G thus giving a one-parameter group in G. If a is rational the image is a simple closed curve but if a is irrational the image is everywhere dense on the torus.

1.14. *Products of closed sets*

If A and B are subgroups of a group G, AB is not necessarily a subgroup. However if A is an *invariant* subgroup (that is $g^{-1}Ag = A$, $g \in G$) and B is a subgroup then AB is a subgroup.

The product of closed subsets, even if they are subgroups, need not be closed. As an example let G be the additive group of real numbers, H_1 the subgroup of integers $\{\pm n\}$ and H_2 the subgroup $\{\pm n\sqrt{2}\}$. The product H_1H_2 is countable and a subgroup but it is not a closed set.

It will be shown later that if A is a compact invariant subgroup and B is a closed subgroup then AB is a closed subgroup (corollary of 2.1).

1.15 Neighborhoods of the identity

Let G be a topological group and U an open subset containing the identity e. We showed in 1.13 that xU is open and clearly $x \in xU$. Conversely if $x \in O$, O open, then $U = x^{-1}O$ is an open set containing e.

If a collection of open sets $\{U_a\}$ is a basis for open sets at e then every open set of G is a union of open sets of the form $x_a U_a$, $x_a \in G$, $U_a \in \{U_a\}$ and the topology of G is completely determined by the basis at e. In particular the collection $\{xU_a\}$ is a basis for open sets at x (so also is $\{U_a x\}$).

If U is a neighborhood of e, U^{-1} is a neighborhood of e and $U \cap U^{-1}$ is a *symmetric* neighborhood of e.

THEOREM. *Let G be a topological group and U a neighborhood of e. There exists a symmetric neighborhood W of e such that $W^2 \subset U$.*

Since $e \cdot e = e$ and the product is simultaneously continuous in x and y there must exist neighborhoods V_1 and V_2 of e such that $V_1 V_2 \subset U$. Define $W = V_1 \cap V_1^{-1} \cap V_2 \cap V_2^{-1}$. Then $W^2 \subset U$, and this completes the proof.

COROLLARY. *Let G be a topological group. If $x \neq e$, there exists a neighborhood W of e such that $W \cap xW$ is empty.*

Since G is a topological space there is a neighborhood V_1 of e not containing x or there is a neighborhood V_2 of x not containing e. In the second alternative, xV_2^{-1} is a neighborhood of e not containing x. In either case there is a neighborhood U of e not containing x. Let W be a symmetric neighborhood of e, $W^2 \subset U$. If $W \cap xW$ were not vacuous there would exist w, $w' \in W$ with $w' = xw$. But this gives $x = w'w^{-1} \in W^2 \subset U$ which is false.

1.15.1.   If G is a topological group then a set S of open neighborhoods $\{V\}$ which forms a basis at the identity e has the following properties:

   a)   the intersection of all V in S is $\{e\}$
   b)   the intersection of two sets of S contains a third set of S
   c)   given U in S there is a V in S such that $VV^{-1} \subset U$
   d)   if U is in S and $a \in U$ then there is a V in S such that $Va \subset U$

e) if U is in S and a is in G there is a V in S such that $aVa^{-1} \subset U$.

Conversely a system of subsets of an abstract group having these properties may be used to determine a topology in G as will be formulated in the following theorem, the proof of which is contained in main part in remarks already made.

THEOREM. *Let G be an abstract group in which there is given a system S of subsets satisfying a) to e) above. If open sets in G are defined as unions of sets of the form Va, $a \in G$ then G becomes topological with S a basis for open sets at e. This is the only topology making G a topological group with S a basis at e.*

1.16. Coset spaces

Let G be a group and H a subgroup. The sets xH and yH, x, $y \in G$, either coincide or are mutually exclusive; and $xH = yH$ if and only if $x^{-1}y \in H$. Each set xH is called a *coset* of H, more specifically a left coset. Right cosets Hx, Hy will be used infrequently. We use the notation G/H for the set of left cosets. When G is a space G/H will be made into a space (see below 1.16.2) but we speak of it as a space in the present case also, although it carries no topology at present.

If H is an invariant subgroup, that is if $x^{-1}Hx = H$, equivalently if $Hx = xH$, $x \in G$, then

$$xHyH = xyH, \qquad x, y \in G$$

is a true equation in sets of elements of G. In this case the coset space becomes a group, the factor group G/H.

1.16.1 DEFINITION: *By the natural map T of a group G onto the coset space G/H, H being a subgroup of G, we mean the map*

$$T : x \to xH, \qquad x \in G, \ xH \in G/H.$$

For any subset $U \subset G$ we have

$$T^{-1}(T(U)) = UH \subset G.$$

Let G be a topological group, H a subgroup. It is useful to topologize the coset space and to do this so that the natural map T

is continuous. It will become clear as we proceed that unless the group H is a closed subgroup of G it will not be possible in general to have T continuous and G/H a topological space; for this reason only the case where H is closed will be considered.

1.16.2. DEFINITION. *Let G be a topological group and let H be a closed subgroup of G, that is a subgroup which is a closed set. By an open set in G/H we mean a set whose inverse under the natural map T is an open set in G.*

THEOREM. *With open sets defined as above, G/H is a topological space and the map T is continuous and open. If $xH \neq yH$, there exist neighborhoods W_1 and W_2 of xH and yH respectively such that $W_1 \cap W_2$ is empty.*

Let U denote an arbitrary open set in G. Then UH is open (1.13) and is the inverse of $T(U)$. Since $T(U)$ is open in G/H, T is an open map.

Let x, $y \in G$ and $xH \neq yH$. Then $x \notin yH$, and yH is closed because H is closed. There exists a neighborhood U of e such that $Ux \cap yH$ is empty. Let W be open, $e \in W$, and $W^2 \subset U$ (Theorem in 1.15). Then if $WxH \cap WyH$ is not empty we can find w, $w_1 \in W$ and h, $h_1 \in H$ so that $w_1 xh = wyh_1$. But this leads to $w^{-1}w_1 x = yh_1 h^{-1}$ and implies that Ux meets yH. Therefore $WxH \cap WyH$ is empty. The sets Wx and Wy are open. Therefore $W_1 = T(Wx)$ and $W_2 = T(Wy)$ are open in G/H; $xH \in W_1$, $yH \in W_2$, and $W_1 \cap W_2$ is empty. This is the main part of the proof and depends on the fact that H is closed. We have proved more than condition 4) of 1.1. The remaining conditions of 1.1 are easy to verify. The fact that T is a continuous map is stated in the definition of open set in G/H. The fact that T is open was proved in the first paragraph.

COROLLARY. *If G is a topological group, x, $y \in G$, $x \neq y$, then there exist neighborhoods W_1 of x and W_2 of y such that $W_1 \cap W_2$ is empty.*

To see this it is only necessary to take $H = e$. A space in which every pair of distinct points belong to mutually exclusive open sets is called a *Hausdorff space*. Therefore it has been shown that a topological group G and a coset-space G/H, H closed in G, are Hausdorff spaces.

COROLLARY. *Suppose that G is a topological group and H a closed invariant subgroup. Then with the customary definition of product: $(xH)(yH) = xyH$, G/H becomes a topological group. The natural map of G onto G/H is a continuous and open homomorphism.*

1.17. A family of neighborhoods

Suppose that we are given a topological group G and a sequence of neighborhoods of e: $Q_0, Q_1, \ldots$. By repeated use of the theorem of 1.15 we can choose a sequence of symmetric open neighborhoods of e: $U_0, U_1, \ldots$, with $U_0 = Q_0$, such that

1) $\quad U_{n+1}^2 \subset U_n \cap Q_n, \quad n = 0, 1, \ldots,$.

In this section we shall show how to imbed the sets U_n in a larger family of neighborhoods possessing a multiplicative property which generalizes 1). We shall use this family in the next section to construct a real and non-constant function which is continuous on G. In 1.22 we shall use a similar family in order to construct a metric in a metrizable group. We remark in passing that in groups which do not satisfy the first countability axiom the set: $\cap U_n$ may be of considerable interest (see 2.6); it is a closed group (if $x, y \varepsilon \cap U_n$ then for every n, $xy \varepsilon U_{n+1}^2 \subset U_n$).

Now, for each dyadic rational $r = k/2^n$, for $n = 0, 1, \ldots$ and $k = 1, \ldots, 2^n$ we define an open neighborhood V_r of e as follows:

2) $\quad V_{1/2^n} = U_n$, for all n,

and then using 3) and 4) alternately by induction on k,

3) $\quad V_{2k/2^{n+1}} = V_{k/2^n}$

4) $\quad V_{(2k+1)/2^{n+1}} = V_{1/2^{n+1}} V_{k/2^n}$.

Each V_r depends on the dyadic rational r, and not on the particular representation by $k/2^n$. The entire family has the property

5) $\quad V_{1/2^n} V_{m/2^n} \subset V_{m+1/2^n}, \quad m + 1 \leqq 2^n$.

For $m = 2k$, 5) is an immediate consequence of 3) and 4). For $m = 2k + 1$, the left side of 5) becomes

$$V_{1/2^n}(V_{1/2^n}V_{k/2^{n-1}}) \subset V_{1/2^{n-1}}V_{k/2^{n-1}}.$$

The right side of 5) becomes $V_{(k+1)/2^{n-1}}$. This sets up an induction on n and since 5) holds for $n = 1$, we have proved that 5) is true for all n. It follows from 5) and also more directly that:

6) $V_r \subset V_{r'}$ if $r < r' \leq 1$.

1.18. Complete regularity

THEOREM. *Suppose that G is a topological group and that F is a closed subset of G not containing e. Then one can define on G a continuous real function f, $0 \leq f(x) \leq 1$, $f(e) = 0$, $f(x) = 1$ if $x \varepsilon F$.*

In virtue of the property described in the theorem, G is called a *completely regular space at the point e*. The theorem is due to Pontrjagin (see A. Weil [1]).

Set $Q_0 = G - F$ and (setting $Q_n = Q_0$, for every n) construct a family of sets V_r as in the preceding section.

Define $f(x)$, $x \varepsilon G$, as follows:

1) $f(x) = 0$ if $x \varepsilon V_r$ for every r,

2) $f(x) = 1$ if $x \notin V_1$,

and in all other cases

3) $f(x) = \text{l.u.b.}_r\{r \leq 1, x \notin V_r\}$.

It is clear that e belongs to every V_r and that $F = G - V_1$ and does not meet V_1, so that there remains only to prove that f is continuous; let $\epsilon > 0$ be given and let n be a positive integer such that $1/2^n < \epsilon$.

Now suppose that $f(x) < 1$ at some point $x \varepsilon G$. Then there is a pair of integers m and k such that $k > n$ (same n as above), $m < 2^k$ and (interpreting V_0, if it occurs, as the null set),

$$x \varepsilon V_{m/2^k} - V_{m-1/2^k}.$$

Let y be an arbitrary element in the neighborhood $V_{1/2^k}x$. Then

$$y \,\varepsilon\, V_{1/2^k} V_{m/2^k} \subset V_{(m+1)/2^k}.$$

By the choice of y, $yx^{-1} \varepsilon V_{1/2^k}$; therefore $xy^{-1} \varepsilon V_{1/2^k}$ and $x \varepsilon V_{1/2^k}y$. It follows from this that y cannot belong to $V_{(m-2)/2^k}$ and this shows that

$$(m-2)/2^k \leqq f(y) \leqq (m+1)/2^k.$$

Concerning x we know that

$$(m-1)/2^k \leqq f(x) \leqq m/2^k$$

and we may conclude from both inequalities that

$$|f(x) - f(y)| \leqq 2/2^k \leqq 1/2^n < \epsilon.$$

Suppose next that $f(x) = 1$, and choose $k > n$ as before. Let y be an arbitrary element in $V_{1/2^k}x$. Now y cannot belong to $V_{m/2^k}$ with $m < 2^k - 2$ without implying that $f(x) < 1$. It follows that

$$1 - 2/2^k \leqq f(y) \leqq 1$$

and again we get $|f(x) - f(y)| \leqq \epsilon$. This concludes the proof that $f(x)$ is continuous on G.

COROLLARY. *A topological group is completely regular at every point.*

1.19. Homogeneous spaces

THEOREM: *Let G be a topological group, H a closed subgroup. Each element of G determines a homeomorphism of the coset space G/H onto itself, and G becomes a group of homeomorphisms of this space; furthermore G is transitive on the space: i.e., each point may be carried to any other by an element of G.*

Let $a \,\epsilon\, G$. Associate to a the mapping

$$T_a : xH \to axH.$$

This is a one-one transformation of G/H onto itself with $T_{a^{-1}}$ as inverse. These transformations are open and each transformation is a homeomorphism.

Since $T_b T_a$ is given by

$$T_b T_a(xH) = T_b(axH) = baxH = T_{ba}(xH),$$

the association of $a \in G$ and T_a makes G a group of transformations of G/H. It is clear that xH is carried to yH by T_a with $a = yx^{-1}$. This completes the proof.

A space is called *homogeneous* when a group of homeomorphisms is transitive on it. We have shown that G/H is homogeneous with G being the transitive group of homeomorphisms. This implies that the unit segment R_1, for example, cannot be the underlying space of a group or even a coset-space G/H since an end-point of R_1 cannot go to an interior point by a homeomorphism of R_1.

It follows from the simultaneous continuity of $ax \in G$ in the factors a and x that the image point axH of xH under T_a is continuous simultaneously in the counter-point xH and the element T_a of G. This makes G an instance of what is called a topological transformation group of a space M, which will be defined below.

However, we shall be principally concerned with transformation groups which are locally compact and separable, acting on spaces which are topologically locally euclidean.

1.20. Local groups

An open neighborhood of the identity of a topological group when it is regarded as a space in the relative topology has some of the properties of a group. There will usually be pairs of elements for which no product element exists in the neighborhood. A structure of this kind is called a local group and will be defined below. Local groups often arise in a natural way, especially in the case of analytic groups (Lie groups of transformations) and they have been intensively studied in that form.

DEFINITION. A space G is called a *local group* if a product xy is defined as an element in G for some pairs x, y in G and the following conditions are satisfied.

1) There is a unique element e in G such that ex and xe are defined for each x in G and $ex = xe = x$.

2) If x, y are in G and xy exists then there is a neighborhood U of x and a neighborhood V of y such that if $x' \epsilon U$, $y' \epsilon V$ then $x'y'$ exists. The product xy is continuous wherever defined.

3) The associative law holds whenever it has meaning.

4) If $ab = e$ then $ba = e$. An element b satisfying this relation is called an inverse and is denoted by a^{-1}. We assume that a^{-1} is unique and continuous where defined, and that if it exists for an element a it exists for all elements in some neighborhood of a. Note that a^{-1} always exists in some neighborhood of e. In fact there exists a symmetric open neighborhood U of e such that U^2 is defined.

The above definition is somewhat redundant.

EXAMPLE. Any neighborhood O of the identity in a topological group is a local group if the neighborhood is open.

We shall call two local groups *isomorphic* if there is a homeomorphism between their elements which carries inverse to inverse and product to product in so far as these are defined. However, in some applications it is natural to regard two local groups as equivalent if they belong to the same local equivalence class; that is, a neighborhood of e in one is isomorphic to a neighborhood of e in the other. In this book an isomorphism means a one-one correspondence which is a homeomorphism and preserves group operations so far as they are defined.

LEMMA. *Let G be a local group with U the symmetric open neighborhood of e described in the definition. Given any neighborhood V of e, $V \subset U$, there is a symmetric neighborhood W of e, $W^3 \subset V$. The product sets AB, BC, $(AB)C$, $A(BC)$ exist for A, B, $C \subset W$ and $(AB)C = A(BC)$. The set AB is open if either A or B is open. The sets A, bA, and Ab are homeomorphic for $b \epsilon W$. Any two points of W have homeomorphic neighborhoods.*

The proof is omitted, the details being as in 1.13, 1.14, 1.15.

1.21. Coset space of local groups

DEFINITION. *A closed subset H of a local group G is called a subgroup if a neighborhood V of e exists such that V^2 is defined, V open and symmetric, and*:

1) $x \in H \cap V$ *implies* $x^{-1} \in H \cap V$

2) $x, y \in H \cap V$ *implies* $xy \in H$.

The subgroup is called *invariant* if a V exists satisfying 1) and 2) and also

3) $y \in V$, $h \in H \cap V$ *implies* $y^{-1}hy \in H$.

LEMMA. *Let G be a local group, H a subgroup, V a neighborhood as described above, and let W be a symmetric open neighborhood of e with $W^{16} \subset V$. Then for x, y $\in$ W, $x^{-1}y \in H$ is an equivalence relation and $x^{-1}y \in H$ if and only if $xH \cap W = yH \cap W$.*

The relation $x^{-1}y \in H$ is reflexive since $x^{-1}x = e \in H$. It is symmetric since if $x, y \in W$ then $x^{-1}y \in W^2$; hence $y^{-1}x$ is defined and belongs to W^2. Now $x^{-1}y \in H$ implies $y^{-1}x \in H$. The relation is transitive because $x, y, z \in W$, $x^{-1}y$ and $y^{-1}z \in H$ implies $x^{-1} z \in H$ by condition 2).

It can be seen that the sets $xH \cap W$, $x \in W$ are the equivalence classes of this equivalence relation. They are called the *local cosets* and are the points of a coset space $W/(H \cap W)$. We shall now define the topology for this space.

Let T denote the *natural map* of W onto $W/(H \cap W)$ defined by

$$T : x \rightarrow (xH \cap W).$$

Note that T (for local groups) is defined only on W. Let the open sets of the coset space be those which have an open set in W as inverse under T. Then T is open and continuous.

COROLLARY. *With the same assumptions and notation as in the preceding Lemma let H be invariant. Then $W/(H \cap W)$ is a local group when the product is defined in the natural way.*

The product function is defined for $x, y \in W$ by:

$$(xH \cap W)(yH \cap W) = (xyH \cap W).$$

To prove that this is single-valued on cosets it needs to be shown that for h, $h' \in H$ and xh, $yh' \in W$, the product $(xy)^{-1}(xhyh')$ exists and is in H. The product is in W^4 and defines an element of V. This element can be seen to be in $H \cap V$ by the use of the associative law and condition 3) above; it must be verified that all indicated products are defined. Finally let U_0 be an open neighborhood of e in G, $U_0^2 \subset W$. It can be seen that the product of each pair of elements in the open set $T(U_0 H \cap W)$ is defined as an element of $W/(H \cap W)$. The remaining details can be verified by the reader.

THEOREM. *A local group G possesses such a family of neighborhoods as is defined in 1.17. For every subgroup H of G and neighborhood W of e (this neighborhood chosen as above) the coset space $W/(W \cap H)$ is a Hausdorff space.*

The proof is the same as in 1.16 and 1.17.

Some of the principal results of the succeeding chapters are valid for local groups as well as global groups. However the consideration of local groups in each preliminary Lemma and Theorem is not feasible in a work of this kind. In the sequel we shall only occasionally need to make explicit mention of the local groups.

1.22. Invariant metric

The construction of metrics and invariant metrics in groups was carried out by Garrett Birkhoff [3] and Kakutani [1] independently.

A local group is called metric if some neighborhood of the identity is metric.

THEOREM. *Let G be a topological group whose open sets at e have a countable basis. Then G is metrizable (1.9) and, moreover, there exists a metric which is right invariant.*

Let U_i, $i = 1, 2, \ldots$, be a countable basis for open sets at e and for each positive integer n let $O_n = \cap_1^n U_i$. The sets O_n are monotonic decreasing and form a basis at e. Let V_r, for each dyadic rational $r = k/2^n$, $1 \leq k \leq 2^n$, $n \in I$, be a family of neighborhoods of e in G such as is constructed in 1.17. Define a function $f(x, y)$ on G as follows:

1) $f(x, y) = 0$ if and only if $y \in V_r V_r^{-1} x$ for every r,
2) otherwise $f(x, y) = $ l.u.b. $\{r; y \notin V_r V_r^{-1} x\}$.

For any set U and element a in G, $y \in Ux$ if and only if $ya \in Uxa$. From this it follows that $f(x, y)$ is right invariant:

3) $f(x, y) = f(xa, ya)$.

Next from the fact that $V_r V_r^{-1}$ is symmetric it follows that $xy^{-1} \in V_r V_r^{-1}$ if and only if $yx^{-1} \in V_r V_r^{-1}$ for the same r. It follows that f is symmetric

$$f(x, y) = f(y, x).$$

The sets $V_{1/2^n}$ are symmetric so that $V_{1/2^n} V_{1/2^n}^{-1} = V_{1/2^n}^2 \subset V_{1/2^{n-1}}$ by 5) of 1.17. Also $V_{1/2^{n-1}} \subset O_{n-1}$ and $\cap O_{n-1} = e$. It follows now that

$$f(x, y) = 0 \text{ if and only if } x = y.$$

We next define the distance function:

*) $d(x, y) = $ l.u.b.$_u$ $| f(x, u) - f(y, u) |$.

The definition shows that $d(x, y) = d(y, x)$,

$$d(x, y) \geqq f(x, y) \geqq 0, \text{ and } d(x, x) = 0.$$

If $d(x, y) = 0$ then $f(x, y) = 0$ and $x = y$. Finally the triangle inequality:

$$d(x, z) = \text{l.u.b.}_u | f(x, u) - f(y, u) + f(y, u) - f(z, u) |$$
$$\leqq d(x, y) + d(y, z).$$

The right invariance of the metric is shown as follows:

$$d(xa, ya) = \text{l.u.b.}_u | f(xa, u) - f(ya, u) |$$
$$= \text{l.u.b.}_{va} | f(xa, va) - f(ya, va) |$$
$$= \text{l.u.b.}_v | f(x, v) - f(y, v) |$$
$$= d(x, y).$$

Finally we must show the equivalence of the original neighborhoods of G and the sphere neighborhoods of the metric. It is sufficient to verify this at e because of the invariance of the metric on the one hand and the translation properties of G (1.13) on the other. Let $S_{1/2^n}$ denote the set of elements whose distance from e is $< 1/2^n$. The fact that

$$V_{1/2^{n+1}} \subset S_{1/2^n}$$

shows that the metric spheres are neighborhoods of e in the original topology. It remains to show that they form a basis at e. Let a neighborhood O of e be given. Since the sets O_n are a basis at e there is an integer k such that $O_k \subset O$. Then if $x \in S_{1/2^{k+1}}$, $d(e, x) < 1/2^{k+1}$ and $f(e, x) < 1/2^{k+1}$, and finally

$$x \in V_{1/2^{k+1}}^2 \subset V_{1/2^k} \subset O_k, \text{ hence}$$
$$S_{1/2^{k+1}} \subset O_k.$$

This completes the proof.

1.23. Metric homogeneous space

THEOREM. *Let H be a closed subgroup of a metric group G. Then G/H is metrizable.*

Let $d(x, y)$ be a right invariant metric in G. Define $D(xH, yH)$ as follows:

$$D(xH, yH) = \text{g.l.b. } d(xa, yb), \ a, \ b \in H.$$

Then for every $\epsilon > 0$, H contains elements a, b, c, d such that

$$D(xH, yH) + D(yH, zH) + 2\epsilon$$
$$\geqq d(xa, yb) + d(yc, zd)$$
$$= d(xa, yb) + d(yb, zdc^{-1}b)$$
$$\geqq d(xa, zdc^{-1}b)$$
$$\geqq D(xH, zH).$$

This shows that D satisfies the triangle inequality. It can be verified that D is single-valued in $G/H \times G/H$, that D is symmetric, and that $D(xH, yH) = 0$ if and only if $xH = yH$. Then D is a metric.

The set yH belongs to the set $S_{1/2^n}(e) \cdot xH$ if $D(xH, yH) < 1/2^n$. Therefore any neighborhood of xH in the original topology includes a metric neighborhood. The converse can also be shown. Therefore the metric D induces the topology of G/H.

1.24. Connectedness

A subset M of a topological space S is called *connected* if whenever $M = A \cup B$, where A and B are open relative to M and are not empty, it follows that $A \cap B$ is not empty.

LEMMA. *If M is connected and A is a subset which is both open and closed relative to M, than $A = M$, or A is the empty set.*
This follows from the definition.

LEMMA. *If G is a topological group and H is a subgroup which is open, then H is also closed.*
Since H is open each coset of H is open. The complement of H is the union of cosets of H. Hence the complement of H is open so H must be closed.

THEOREM. *If G is a connected group and W is an open neighborhood of e, then $G = \cup_n W^n$.*
The set $H = \cup_n (W \cap W^{-1})^n$ is an open subgroup of G. Therefore H is a closed subgroup, and then it follows from the connectedness of G that $H = G$.

A union of an arbitrary number of connected subsets is connected, provided every two of the sets have a point in common. It follows that to each point $x \in S$ there is uniquely associated a *maximal connected subset* of M containing x; these maximal connected subsets are called *components* of S. The closure of a component is connected — therefore the component is closed. A space is called *totally-disconnected* when each point is a component.

LEMMA. *If M is connected and f is a continuous map of M into a space N then $f(M)$ is connected.*
$$\text{If } f(M) = A \cup B \text{ and } A \cap B = \Phi,$$
then $M = f^{-1}(A) \cup f^{-1}(B)$ and $f^{-1}(A) \cap f^{-1}(B) = \Phi$.

COROLLARY. *If G is a connected group and H is a closed subgroup, then G/H is a connected space.*

DEFINITION. *A space is called normal if any two disjoined closed sets are contained in disjoined open sets; two sets are disjoined (disjunct, mutually separated) if they have no common points.*

LEMMA. *A compact Hausdorff space is normal.*

The proof is straightforward and is omitted.

LEMMA. *Let S be a compact Hausdorff space and x a point in S. Let Q be a set of indices $\{q\}$. If $\{A_q\}$ is the set of all compact open subsets containing x, then $C = \cap_q A_q$ is the x-component of S.*

Let

$$\cap_q A_q = X \cup Y, \quad X \cap Y = \Phi,$$

X and Y closed. There exist open sets $V \supset X$ and $W \supset Y$ such that $V \cap W = \Phi$. Hence

$$*) \quad (S - (V \cup W)) \cap (\cap_q A_q) = \Phi$$

and there is a *finite* set of indices $Q' \subset Q$ such that $*)$ continues to hold if the intersection $\cap_q$ is restricted to $q \in Q'$. Let $A = \cap_{q'} A_{q'}$, $q' \in Q'$. Then A is compact and open and $x \in A$.

Finally, since

$$A = (A \cap V) \cup (A \cap W)$$

it follows that $A \cap V$ and $A \cap W$ are compact open sets *only one of which can contain x*. Since the system $\{A_q\}$, $q \in Q$, is maximal, one of $A \cap V$ and $A \cap W$ is in $\{A_q\}$, say $A \cap V \in \{A_q\}$ and therefore $\cap_q A_q \subset A \cap V \subset V$; hence Y is empty.

This shows that $C = \cap_q A_q$ is connected. It is now clear that C is the component containing x.

COROLLARY. *Let S be a compact Hausdorff space, C be a component of S, let F be a closed set and suppose that $F \cap C = \Phi$. Then there is a compact open set A', $C \subset A'$, $F \cap A' = \Phi$.*

It follows from the hypothesis that there is a finite set $Q'' \subset Q$ such that

$$F \cap (\cap_{q''} A_{q''}) = \Phi, \quad q'' \in Q''.$$

Let

$$A' = \cap_{q''} A_{q''}.$$

This is the set required by the corollary.

1.25. The identity component G_0

THEOREM. *Let G be a topological group and G_0 the component of G containing the identity. Then G_0 is a closed invariant subgroup and the factor group G/G_0 is totally disconnected.*

If M is a connected subset of G, M^{-1} is connected, Mx and xM, $x \in G$, are connected (1.13). It follows that the identity component G_0 is a group, and that the cosets xG_0 are also components. We have already remarked that components are closed sets. Then G_0 is a closed subgroup. Since $x^{-1}G_0x$ is connected and contains the identity e for every $x \in G$, it is clear that G_0 is invariant.

Now let M denote a connected subset of G/G_0. Let T be the natural map of G onto G/G_0. We shall show that $T^{-1}(M)$ is connected, thus proving that it is in a coset of G/G_0. Suppose that

$$T^{-1}(M) = A \cup B$$

where A and B are relatively open (and closed) in $T^{-1}(M)$ and $A \cap B$ is empty. Then $T(T^{-1}(M)) = M = T(A) \cup T(B)$. Since there exists an open set, say U, in G, such that $A = U \cap T^{-1}(M)$ and since T is an open map, the set $T(A)$ is open relatively to M. This is true also of $T(B)$. Since M is connected it follows that $T(A) \cap T(B)$ is not empty. Let $xG_0 \in T(A) \cap T(B)$. Since xG_0 is a connected subset of G,

$$xG_0 = (xG_0 \cap A) \cup (xG_0 \cap B)$$

leads to a contradiction.

A subgroup of a group G is called *central* if each of its elements commutes with every element in the whole group. Subgroups H and $g^{-1}Hg$ are called *conjugate*.

THEOREM. *If G is a connected group and H is an invariant totally disconnected subgroup then H is central.*

Let x be an element of H and consider the map of G into H depending on x:

$$g \to gxg^{-1}, \ g \, \varepsilon \, G.$$

Since H contains no connected set with more than one point, the image of G which contains x, must coincide with x. This completes the proof.

The set of all central elements forms a group called the *center*.

1.26. Transformation groups

We shall soon confine our attention to groups which are locally compact, and we shall be particularly interested in the transformation groups of locally-euclidean spaces. For the time being, we continue to study the more general phenomenon.

DEFINITION 1. *Let M denote a Hausdorff space and G a topological group each element of which is a homeomorphism of M onto itself:*

$$(1) \quad f(g;\ x) = g(x) = x' \in M;\ g \in G,\ x \in M.$$

The pair (G, M) or, sometimes, G itself, will be called a topological transformation group if for every pair of elements of G, and every x of M,

$$(2) \quad g_1(g_2(x)) = (g_1 g_2)(x),$$

and if $x' = g(x) = f(g;\ x)$ is continuous simultaneously in $x \in M$ and $g \in G$.

From 2) and from the fact that each g is one-one on M, it follows that for every $x \in M$

$$(3) \quad e(x) = x;\ e \text{ the identity in } G.$$

If e is the only element in G which leaves all of M fixed, i.e. if e is the only element satisfying 3) for all x, then G is called *effective*.

DEFINITION 1'. *A pair (G, M) will be called a local transformation group if all conditions of the preceding Definition are fulfilled excepting only that G is assumed to be a local group, and condition 2) holds whenever $g_1 g_2$ is defined.*

Some of the remarks below apply both to local and global transformation groups but we confine attention to the global case for the most part, except when noted.

Let G be a transformation group, $g \in G$, x, $y \in M$ and suppose that

$$g(x) = y \neq x.$$

Since M is a Hausdorff space, there is a neighborhood Y of y not containing x. By the definition of transformation group, there is a neighborhood U of g such that for $g' \in U$, $g'(x) \in Y$. Therefore $g'(x) \neq x$, and it follows that the set of elements of G which leave x fixed is closed. Therefore this set is a closed subgroup of G. We shall denote it by G_x. Similarly, the set of elements of G leaving fixed every point of M is a closed subgroup G_M.

THEOREM. *Let (G, M) be a transformation group, and let K be the closed subgroup of G leaving all of M fixed. Then K is invariant and G/K is an effective transformation group of M under the action*:

$$T^* : (gK)(x) = g(x); \ g \in G.$$

For $x \in M$, $h \in K$, $g \in G$, we must always have: $(g^{-1}hg)(x) = x$. This shows the invariance of K. If $(gK)x = y$ and a neighborhood Y of y is given, there is a neighborhood U of g and X of x such that $g' \in U$, $x' \in X$ imply $g'(x') \in Y$. But then $(g'K)(x') \in Y$. From this it is clear that T^* is continuous simultaneously in gK and x. It can be seen that

$$g_1K(g_2K(x)) = g_1g_2K(x).$$

It follows also that $gK(x) = x$ for every x implies $g \in K$, equivalently $gK = K$.

1.26.1.   If G is a transformation group of M and h is a homeomorphism of M onto itself, then the homeomorphisms

$$\{hgh^{-1}\}$$

of M onto itself form a transformation group which is said to be topologically equivalent to G.

1.26.2.   A topological group G can be regarded as a transformation group on itself as space in several ways, in particular by associating with $a \in G$

1) $a(g) = ag$ (left-translation)
2) $a(g) = ga^{-1}$ (right-translation)
3) $a(g) = aga^{-1}$ (conjugation, taking of transforms).

Also G is a transformation group of a left coset space G/H, H a closed subgroup, by

4) $a(gH) = agH$.

In cases 1) and 2) G is effective. In 3) let Z denote the center of G; then G/Z is effective. In case 4) if

$$agH = gH$$

for every $g \in G$ then

$$a \in gHg^{-1}$$

for every $g \in G$. Then $K = \cap \, gHg^{-1}$ is an invariant subgroup of G, and G/K is effective on G/H. Of course K is a subgroup of H which depends on H as well as G and which may be trivial.

1.26.3. Further examples of transformation groups are given below, proofs are omitted.

1) Let $G = Gl(n, R)$ be the real $n \times n$ matrices $a = (a_{ij})$ with $|a_{ij}| \neq 0$. Let E_n be the space of n real variables $u_1, \ldots, u_n$. Then G is a transformation group of E_n whose elements are

$$T_a^* : a(u) = u'$$

where $u_i' = \Sigma a_{ij} u_j$.

2) With G as above let S^{n-1} in E_n be the $(n - 1)$-sphere defined by $\Sigma u_i^2 = 1$. Let G act on S^{n-1} as follows:

$$T_a'' : a(u) = u''$$

where $u_i'' = u_i'/(\Sigma u_i'^2)^{1/2}$ and u_i' is as above.

Here the effective group is G/K_n, where K_n consists of scalar matrices: $(h\delta_{ij})$, where δ_{ij} is the Kronecker delta, and h is positive.

3) Let G be the group of two by two real matrices with determinant one and let it act on E_2 as in 1).

4) Let G be the group of two by two real matrices of determinant one and let G act on itself by inner automorphisms. As

a space G is the product of a circle and a plane. One-parameter groups fill a neighborhood of the identity and in the large are closed sets which are either circles or lines. No two of them cross and they are permuted by G.

5) For fixed integers m, n let G be the circle group acting on E_4 as follows

$$x_1^1 = x_1 \cos 2\pi mt + x_2 \sin 2\pi mt$$
$$x_2^1 = - x_1 \sin 2\pi mt + x_2 \cos 2\pi mt$$
$$x_3^1 = x_3 \cos 2\pi nt + x_4 \sin 2\pi nt$$
$$x_4^1 = - x_3 \sin 2\pi nt + x_4 \cos 2\pi nt.$$

This can also be viewed as a transformation group on the unit sphere S^3 in E^4 since it leaves S^3 invariant. It is known that the simple closed curves swept out by points of S^3 are linked.

6) A *quasi-rotation* in E_3 in a fixed cylindrical coordinate system (z, r, θ), is a group of homeomorphisms depending on a positive continuous function $F(r, z)$, $0 < r < \infty$, $- \infty < z < \infty$, F bounded on compact sets in E_3 and given by:

$$* \quad h_t : (z, r, \theta) \to (z, r, \theta + 2\pi F(r, z)t),$$

for all real t.

Each point which is not a fixed point moves in a circle about the z-axis. The period of a moving point, i.e. the least positive t for which it is left fixed by *) varies continuously. If h is an arbitrary homeomorphism of E_3 upon itself, then $\{h^{-1}h_t h\}$ defines a topological quasi-rotation-group. As we shall mention later, these groups can be characterized abstractly.

1.26.4. The transformation group G is called *transitive* on M if for every x, $y \in M$ there is at least one $g \in G$, such that $g(x) = y$. As remarked earlier every topological group G is transitive on G/H, H a closed subgroup.

THEOREM. *Let (G, M) be a topological transformation group which is transitive on M. Then the groups of stability G_x, $x \in M$, are conjugate and for any one of them G/G_x is mapped in a con-*

tinuous one-one way onto M by the map

$$T_1 : gG_x \to g(x).$$

Let $x, y \in M$ be given with $g(x) = y$. If $g'(x) = x$, $gg'g^{-1}(y) = y$. It follows that G_x and G_y are conjugate.

Now let x be fixed. If $g'(x) = x$, $gg'(x) = g(x)$ for every $g \in G$ and this shows that the map

$$T_1 : G/G_x \to M$$

defined in the theorem is one-one. It maps G/G_x onto M because G is transitive.

Let T be the natural map $G \to G/G_x$. Then $T_1 T$ maps G onto M:

$$T_1 Tg = g(x).$$

This map is continuous in G by the definition of a transformation group. Let U be open in M. Then $(T_1 T)^{-1} U$ is open in G and $T(T_1 T)^{-1} U$ is open in G/G_x, since T is an open map. This shows that T_1 is continuous as well as one-one. In the cases of most interest it will turn out that T_1 is a homeomorphism.

1.27. Locally euclidean spaces

We have used E_n, $n \in I$, to denote *euclidean n-space*, with real coordinates $x_1, \ldots, x_n$. It is the *topological class* of spaces homeomorphic to E_n which we have in mind, rather than the space endowed with a standard euclidean metric. By the Brouwer Theorem (1911) (see Hurewicz-Wallman) an open subset of E_m and an open subset of E_n cannot be homeomorphic if $m \neq n$. This shows that the possibility of the one-one bicontinuous coordinatization $(x_1, \ldots, x_n)$ of E_n is a topological property. The number n is a topological invariant, and is called the *dimension*.

The term *locally euclidean* is used to describe a topological space E of fixed dimension n each point of which has a neighborhood that is homeomorphic to an open set in E_n. The simplest examples of such spaces are the open subsets of E_n. If a locally euclidean space is connected it is called a *manifold*. For example, the spheres of all dimensions, the ordinary torus, the cylinder, etc., are manifolds.

A locally euclidean space can be covered by a certain number (not necessarily finite) of open sets each homeomorphic to an open set in E_n; let us call such sets, each with a fixed homeomorphism, a *coordinate-neighborhood*. The circle C_1 can be covered by two (or more) coordinate neighborhoods, the two-dimensional sphere S^2 by two or more. However, to describe classical euclidean space one uses the entire family of those coordinate systems which are related to each other by orthogonal transformations. Similarly to define affine n-dimensional space, one uses a larger family, namely all coordinate systems which are affinely related to each other.

To describe a topological manifold one could use in it the family of all possible coordinate neighborhoods. However, if for some purpose a restricted class of coordinate neighborhoods covering the manifold is specified, then one can speak of *admissible* coordinate systems. In general, where two coordinate neighborhoods overlap, the coordinate systems will not be found to be in any simple relation. It may happen that the admissible coordinate systems have been so selected that in every region of overlap of two such systems the two sets of coordinates are related by functions which are differentiable, or analytic.

A manifold is said to be a differentiable manifold and to have a *differentiable structure of class* $C^r (r \geqq 1)$ if there is given a covering family of coordinate neighborhoods in such a way that where any two of the neighborhoods overlap the coordinate transformation in both directions is given by n functions with continuous partial derivatives of order r. A manifold may have essentially different differentiable structures (Milnor). A manifold need not possess a differentiable structure (Kervaire, Smale). The n-sphere (with the possible exception of $n = 3$) has only a finite number of such structures (see Kervaire-Milnor).

In the same way a manifold is said to be a (*real*) *analytic manifold* and to have a (*real*) *analytic structure* if there is given a covering family of coordinate neighborhoods in such a way that where any two overlap the coordinate transformation in both directions is given by n functions which are real analytic, that is in some neighborhood of each point of the overlap they can be expanded in power series.

The definition of *a complex analytic manifold and structure* is similar to the above. Such a manifold of course has an even number of real dimensions; it is automatically real analytic. However, there are many real analytic manifolds of even dimension which cannot be given a complex analytic structure; thus the existence of a complex analytic structure is a much stronger property than the existence of a differentiable or even real analytic structure.

1.27.1. Suppose that M is an n-dimensional manifold and that x and y are points of M belonging to a set U which is homeomorphic to an open n-sphere. Then it is not difficult to describe a homeomorphism of M onto itself which keeps fixed all points of M not inside of U, and which maps x onto y. Using this and using the connectedness of M, and being given an arbitrary pair of points x and y of M one can find a homeomorphism of M onto itself mapping x on y. We shall not give the details.

1.27.2. Suppose that M and M' are manifolds and that there is given a continuous map f of M onto M'. The map f is called a covering map and M is said to cover M' if the following conditions are satisfied

a) for each y in M' there is an open neighborhood V of y such that $f^{-1}(V)$ is the union of disjoined open sets U_x, where there is a U_x for each $x \in f^{-1}(y)$ and $x \in U_x$

b) f is a homeomorphism of U_x onto V for each x in $f^{-1}(y)$. For each $y \in M'$, each $x \in f^{-1}(y)$ is called a *covering point*. It is clear that M and M' are of the same dimension, and it is clear that each point of $f^{-1}(y)$ is an isolated point of $f^{-1}(y)$ (each point is a relatively open subset) so that $f^{-1}(y)$ is a *discrete* set.

By way of example, let M' be the ordinary torus with momentarily convenient coordinates u, $v : 0 \leq u$, $v < 1$, and let M be the ordinary plane with real coordinates x and y. Define the map $f(M) = M'$ by

$$(x, y) \to (u, v) \text{ if and only if } x \equiv u \text{ and } y \equiv v \pmod{1}.$$

This pair of manifolds can also be regarded as an example of a group M covering a factor-group M'. Thus let M now denote the two-dimensional vector space V_2, and let x_1 and y_1 denote two independent vectors in V_2. Let D be the (countable, discrete) subgroup of V_2 consisting of the linear combinations of x_1 and y_1 with integral coefficients. Finally, let M' now denote the toral group V_2/D.

A more general example is the following:

If G is any connected locally euclidean group and H is a discrete subgroup of G then G covers the coset-space G/H under the natural map $G \to G/H$.

A manifold M' is called *simply connected* if whenever it is covered by a manifold M, the covering map $f : f(M) = M'$, is a homeomorphism (in that case $f^{-1}(y)$ is single-valued). Euclidean spaces of all dimensions and the sphere-spaces of dimension *greater than one* are simply connected manifolds; the one-dimensional sphere (circumference of a circle) and more generally, the toral spaces of all dimensions are not simply connected. We shall not really need this concept until the last chapter, and we postpone discussion of it until then.

CHAPTER II

Locally Compact Groups

2.0. Introduction

A topological group is called locally compact if the group-space is locally compact (similarly it is called locally-euclidean or connected or whatever the case may be if the space of the group has the corresponding property). In the case of all local topological properties it is sufficient of course that the group possess the property at some one element, say the identity. Then since group translations are homeomorphisms, the same local property holds at every point. Therefore a group is locally compact if the identity has a compact neighborhood.

The class of locally compact groups will occupy us through this and the next two chapters. By way of a first set of useful examples of these groups consider the following:

1) the group of order two (consisting of two elements) and also the topological product of an infinite number of copies of this group (the group space is the Cantor Discontinuum if the number of copies is countably infinite),

2) the "dyadic" group D_2 whose elements are the formal series in powers of 2:

$$g = a_0 + a_1 2 + \ldots a_n 2^n + \ldots, \text{ each } a_n = 0, 1$$

and elements are added as though they might represent integers, but with (possibly) *infinite carry-over*. Thus

$$g_1 = 1 \text{ and } g_1^{-1} = 1 + 1.2 + \ldots + 1.2^n + \ldots$$

are inverses. The group is abelian; the topology is determined by the following choice of neighborhoods of the zero, or identity:

$$U_n = \{g \varepsilon D_2, \ a_i = 0 \text{ if } i < n\}, \quad n = 1, 2, \ldots$$

so that D_2 is compact and totally disconnected. The space of

D_2 is homeomorphic to the Cantor Discontinuum. We shall call the element g_1 the *generator* of D_2; the "powers" of g_1 are everywhere dense in D_2.

3) the product of an infinite number of groups isomorphic to C_1, the group of reals mod one; this is an "infinite-dimensional" group.

4) the product of D_2 and C_1 or D_2 and V_1, the group of reals. These are locally isomorphic groups, "one-dimensional", but clearly not locally euclidean (see, also, 2.14).

5) the group of rigid-motions of the ordinary plane. This is a locally euclidean group (three-dimensional) and a Lie group (see below).

All discrete groups (the identity element being an open subset) are locally compact, and uninteresting from a topological point of view. The groups C_1 and V_1 as well as 5) are also examples of Lie groups. In appropriate context, the discrete groups may be counted as Lie groups of dimension zero.

The Lie groups constitute the most important class of locally compact groups for two distinct reasons. First, they include all of the most important groups of geometry and analysis (orthogonal groups, affine groups, projective groups, etc.). Second, all locally compact groups may be approximated by Lie groups, roughly speaking. Speaking precisely: let G be a locally compact group. Then G has an *open* subgroup G' and G' has arbitrarily small compact invariant subgroups H such that G'/H is a Lie group.

The proof of this approximation-theorem will run through three chapters, along with other developments. Let us now state briefly what Lie groups are.

DEFINITION. *A topological group is called a (real) Lie group if the identity component is open and if this component may be given an analytic structure (1.27) in which the coordinates of the product element $z = xy$ are analytic functions of the coordinates of the elements x and y, and the coordinates of z^{-1} are analytic functions of the coordinates of z. A complex Lie group is defined similarly.*

The fact that the operation of forming the inverse is analytic follows from the analytic multiplication by the theory of implicit functions, so the definition is slightly redundant.

A complex Lie group is a real Lie group but not necessarily conversely. Clearly, a group isomorphic to a Lie group is a Lie group.

EXAMPLES. The additive group $M_n(R)$ of $n \times n$ real matrices (a_{ij}) has E_{n^2} as underlying space. The components a_{ij} arranged in some order may be taken as analytic coordinates and the formulas of composition are

$$c_{ij} = a_{ij} + b_{ij}.$$

The non-commutative group $Gl(n, R)$ of non-singular $n \times n$ real matrices (a_{ij}) under multiplication has an open subset of E_{n^2} as underlying space. The components may again be taken as analytic coordinates and the law of composition is

$$c_{ij} = \Sigma_k a_{ik} b_{kj}.$$

This multiplication, and the operation of forming the inverse element, are expressed by analytic functions.

It is very useful that the closed subgroups of M_n are easy to find; the closed connected subgroups are themselves linear vector subspaces with $k \leq n^2$ generators. The simplest of these, the one-parameter groups, are given by the set of matrices (ta_{ij}), $-\infty < t < \infty$, a_{ij} any *fixed* real numbers. Through the *exponential map* (2.16) relating M_n to $Gl(n, R)$ it is possible to study the closed subgroups of the multiplicative group. We do this to the extent of proving that they are Lie groups (von Neumann, 1929 [1]). These subgroups, of course, are locally compact.

In 2.20 we prove that compact groups G have arbitrarily small subgroups H such that each G/H is isomorphic to a subgroup of $Gl(n, R)$, first proved by von Neumann [2] in 1933. In 2.21 we study locally compact abelian groups (Pontrjagin [1], 1934). In preparation for all of this, and for the study of locally compact groups in Chapter III, we show the existence of an invariant integral (Haar [1], 1933), and discuss its properties.

Although it is just as easy to set up an integral for the non-separable groups as for the separable, it is interesting to consider the connection between the two cases and we discuss this in 2.19. In the first sections we develop some of the simpler properties of locally compact groups, showing among other things that they can be approximated by separable metric groups (Kakutani-Kodaira [1], 1944). A number of sections are devoted to concepts relating to transformation groups, looking to the final chapters of the book.

2.1. Compact subsets of topological groups

We saw in 1.14 that the product of closed subsets of a group is not necessarily a closed set. However, the product *is* closed if one of the factors is compact. This, and related facts, will now be shown.

LEMMA. *Let G be a topological group, let F be a closed subset and C a compact subset such that $F \cap C = \Phi$. Then there is a neighborhood V of e such that $F \cap CV = \Phi$.*

Let x be in C so that x is in the open set $G - F$. There is a neighborhood W_x of e such that $W_x^2 \subset x^{-1}(G - F)$. There is a set of points $x_i, i = 1, \ldots, n$, and a set of associated neighborhoods W_i such that $x_i(W_i)^2 \subset G - F$, and $\cup_i x_i W_i \supset C$. Set $V = \cap_i W_i$.

Now for any $x \varepsilon C$, $x \varepsilon x_i W_i$ for some i, and $xV \subset x_i W_i^2 \subset G - F$. Then $xV \cap F$ is empty. Clearly $CV \cap F$ is empty, and this completes the proof.

Similarly, there is a V such that $VC \cap F = \Phi$ and there is also a neighborhood U of e, $U^2 \subset V$, and $UC \cap UF = \Phi$.

COROLLARY. *Let C be a compact subset and let B be a closed subset of the topological group G. Then CB and BC are closed.*

Take any y not in BC. Then $B^{-1}y \cap C = \Phi$ and there is a neighborhood V of e such that $B^{-1}yV \cap C = \Phi$. Then $yV \cap BC = \Phi$. Thus the complement of BC is shown to be open. Then BC is closed as asserted in the Corollary. Similarly CB is closed.

The reader will see easily that *a compact subset of a topological group is a normal space* (1.24).

2.1.1. LEMMA. *If B and C are compact subsets of a topological group G, then BC is a compact subset.*

In the product group (1.11.2) $G \times G$, consider the compact set $B' \times C'$ (1.8) where B' is the one-one image of B in the first factor and C' is the one-one image of C in the second factor and therefore B' and C' are compact. Now the product BC in G is the continuous image of $B' \times C'$ by 4) of 1.11 and is therefore compact (1.7).

The Lemma extends to any finite number of factors, the proof is by induction.

2.2. The class of locally compact groups

If G is a locally compact group and if H is a closed subgroup then H is locally compact. The coset-space G/H, say the left-coset space, is also locally compact: for let T be the natural map of G onto G/H, and let U be a compact neighborhood of e in G. Since T is open and continuous, $T(U)$ is a neighborhood of $T(e)$ and $T(U)$ is also compact.

We shall have infrequent occasion to use the following "converse" proposition but we include it for its intrinsic interest. The proof is a modification of one by Gleason [5].

THEOREM. *If a topological group G has a closed subgroup H such that H and the coset-space G/H are locally compact, then G is also locally compact.*

Let T denote the natural map of $G \to G/H$.

Since H is locally compact in the relative topology (1.4) we can find in G a closed neighborhood U of e such that $U \cap H$ is compact. Let U_1 be a closed symmetric neighborhood of e such that $U_1^2 \subset U$. For every x in U_1, $x^{-1}U_1 \cap H$ is a closed subset of $U \cap H$ and is compact. Therefore $x(x^{-1}U_1 \cap H)$ is compact and $U_1 \cap xH$ is compact for every x in U_1. Let U_2 be a closed symmetric neighborhood of e such that $U_2^2 \subset U_1$ and let C denote a compact neighborhood of $T(e)$ contained in $T(U_1)$. Set $V = U_2 \cap T^{-1}(C)$. Now V is a closed neighborhood of e and has the character of a compact collection of compact sets so that one may anticipate that it is compact.

Let $\{O_a\}$ be a collection of *open* sets covering V. We shall prove that there is a finite covering (1.7). Let y be an arbitrary element in C. Pick an element x in $U_1 \cap T^{-1}(y)$; then $T(x) = y$. The family consisting of the open set $G - V$ and the collection $\{O_a\}$ is a covering of $U_1 \cap xH$, and it has a finite subcollection which is also a covering. Let X denote the union of this finite subcollection: $U_1 \cap xH \subset X$; and X is open. Therefore, by the preceding section, there is a neighborhood W of e such that $W(U_1 \cap xH) \subset X$. We may take $W = W^{-1} \subset U_2$. Now let x be fixed; Wx is a neighborhood of x in G and $T(Wx)$ is a neighborhood of y in C.

We now show that for every y' in $C \cap T(Wx)$, $U_2 \cap T^{-1}(y') \subset X$. It is clear that $T^{-1}(y') = wxH$, for some $w \varepsilon W$. If $z \varepsilon U_2 \cap wxH$, then $z = wxh$, for some h in H and $xh = w^{-1}z \varepsilon (U_2)^2 \subset U_1$. Then $xh \varepsilon U_1 \cap xH$ and $z \varepsilon X$. Since $U_2 \cap T^{-1}y'$ is contained in V no point in it can belong to $G - V$. Hence $U_2 \cap T^{-1}(C \cap T(Wx))$ is covered by the finite collection of sets of $\{O_a\}$ which belong to X.

There is a finite set of points x_i, $i = 1, \ldots, n$, and associated neighborhoods W_i, as above, such that the union of the sets $T(W_i x_i)$ covers C. The existence of the desired finite covering for V is now obvious (compare 1.8).

2.3. Open subgroups

We showed in 1.25 that the identity-component of a topological group G is a closed invariant subgroup G_0 such that G/G_0 is totally disconnected. If G is locally compact then G/G_0 is locally compact. In this case, it will next be shown that G/G_0 contains arbitrarily small *compact open* subgroups and corresponding to these G contains open subgroups G' such that G'/G_0 is compact. The groups G' are subgroups which may be generated by a compact neighborhood of the identity, and constitute an important class of the locally compact groups (they are the ones which are actual limits of Lie groups).

Let us recall that if H is an open subgroup of G, then the cosets of G/H are open in G and the complement of H is open, and H is also a closed group. The image of H in G/H being open, it follows that G/H is discrete.

LEMMA. *Let G be a locally compact group and suppose U to be an open and compact neighborhood of e. Then U contains a compact open subgroup.*

Let

$$F = (G - U) \cap U^2,$$

so that F is closed. Since $U \cap F$ is empty there is an open symmetric neighborhood V of e, $V \subset U$, such that $UV \cap F$ is empty; since $UV \subset U^2$, this implies that $UV \subset U$. Then, of course, $UV^n \subset U$ for every n. For this V there follows $\bigcup_n V^n \subset U$. Then $\bigcup_n V^n$ is the desired group; being open it is also closed hence compact.

THEOREM. *Let G be a totally disconnected locally compact group and let V be a compact neighborhood of e. Then V contains a compact open subgroup.*

Let U be an open neighborhood of e, $U^2 \subset V$. Let

$$F = V - U.$$

If F is empty, then $U = V$ is open and compact and the Theorem follows from the preceding Lemma. Suppose that F is not empty. Then, because e is a component of V and V is compact, there exists an open and closed neighborhood W of e, $W \subset V$, such that $W \cap F = \Phi$ (1.24). The preceding Lemma applies to W and since $W \subset U \subset V$ the Theorem is proved.

2.3.1. LEMMA. *Let G be a locally compact group with identity-component G_0. There exists an open subgroup G' of G (necessarily $G' \supset G_0$) such that G'/G_0 is compact. To each such G' there exists a compact neighborhood W such that $G' = W \cdot G_0$; and G' is generated by W ($x' \in G'$ implies $x \in W^n$ for some n).*

Let T denote the natural map

$$G \to G/G_0,$$

and let D be an arbitrary compact open subgroup of G/G_0. Then

$T^{-1}(D)$ is the required G'. Now let T' denote the natural map

$$G' \to G'/G_0.$$

Choose a compact neighborhood W' of e, $W' \subset G'$. Then $T'(W')$ includes a set open in G'/G_0, and G'/G_0 may be covered by a finite set of translates of $T'(W')$, i.e. there is a *finite* subset $F \subset G'$, $e \varepsilon F$, such that

$$G'/G_0 \subset T'(\cup_{g \varepsilon F} W'g).$$

Finally, $W = \cup_{g \varepsilon F} W'g$ is in G' and is compact; $G' = WG_0$ as required.

Since W is a neighborhood of e, W generates an open and closed subgroup of G'. This subgroup contains G_0 because G_0 is connected. Therefore W generates all of G'.

2.4. Intersections of neighborhoods

If U is a neighborhood of e in a locally compact group G and $\{x^{-1}Ux\}$ is a family of transforms of U for x in a set K, then the intersection of members of the family, $\cap x^{-1}Ux$, is not necessarily a neighborhood of e unless K is compact. This special case has important application to groups which are generated by a compact subset.

THEOREM. *Let G be a locally compact group, let K be a compact subset of G and let U be an open neighborhood of e. There exists a neighborhood V of e such that*

$$x^{-1}Vx \subset U, \quad x \varepsilon K.$$

Let x be a point of K. There is a compact neighborhood V_x of e such that $V_x \subset xUx^{-1}$. Then $x^{-1}V_x x \cap F = \Phi$, where $F = G - U$. Therefore there is a compact symmetric neighborhood W_x^0 of e such that $x^{-1}V_x x W_x^0 \cap F$ is empty. This means that $x^{-1}V_x x \cap F W_x^0 = \Phi$. By 2.1 FW_x^0 is closed and so again by 2.1 there is a symmetric W_x' of e such that

$$W_x' x^{-1} V_x x \cap F W_x^0 = \Phi.$$

Let $W_x = W_x^0 \cap W_x'$. Then

$$W_x x^{-1} V_x x W_x \cap F = \Phi.$$

For any $y \, \epsilon \, xW_x$, $y^{-1} \, \epsilon \, W_x x^{-1}$ and

$$y^{-1} V_x y \subset U.$$

Finally there is some *finite* subset X of G such that K is covered by the neighborhoods xW_x as above, $x \, \epsilon \, X$.

Let $V = \cap V_x$, $x \, \epsilon \, X$, where each V_x is associated with W_x as above. Now if $y \, \epsilon \, K$, then $y \, \epsilon \, xW_x$ for some $x \, \epsilon \, X$ and

$$y^{-1} V y \subset W_x x^{-1} V_x x W_x \subset U.$$

This completes the proof.

An equivalent statement of the conclusion of the theorem is that $\cap x^{-1} U x$, $x \, \epsilon \, K$, is a neighborhood of e.

2.5. Totally disconnected groups

A locally compact totally disconnected group contains arbitrarily small compact open subgroups, but none of them are necessarily invariant unless the group is compact.

THEOREM. *If G is a compact totally disconnected group and a neighborhood U of e is given, then there is a compact invariant open subgroup $H \subset U$; G/H is a finite group.*

We saw in 2.3 that U must contain a compact open subgroup, call it H'. Set $H = \cap x^{-1} H' x$, for all x in G. The preceding Theorem shows that H is open and it is clear from its definition that H is invariant and compact. Since G/H is compact and discrete (each coset xH being open), G/H must be finite. This completes the proof.

Suppose now, on the same hypotheses as above, that g is some element of G. Since G/H has only a finite number of cosets, there must be an integer m such that $g^m \, \epsilon \, H$. Therefore almost all elements of the sequence $g^{n!}$ belong to H, $n = 1, 2, \ldots$, and it is natural to say that $g^{n!}$ converges to e even when G is not necessarily separable.

Although compact, totally disconnected groups can be approximated by finite groups, namely the groups G/H, their algebraic structure is not known in the general case *even when the group G*

is abelian. However, the *abelian separable* case has been worked out (See references to Ulm and Zippin in Kaplansky [1]).

EXAMPLE. *There exists a totally disconnected locally compact group G which has no invariant compact open subgroups.*

Let G be the group consisting of an identity element and of all finite and some infinite products (to be described later) of generating elements z and x_n, $n = \ldots, -2, -1, 0, 1, 2, \ldots$. Let the generators be subject to the conditions 1) $x_n^2 = e$ for all n, 2) $x_m x_n = x_n x_m$ for all m and n, 3) $x_m z = z x_{m-1}$ for all m. It follows from 3) that for every i and j, positive or negative, there is an integer n such that 4) $z^{-n} x_i z^n = x_j$.

Let H denote the subset of G consisting of the identity and of all *finite and infinite* products: $x_{n_1} x_{n_2} \ldots$, with $n_1 < n_2 < \ldots$, so that at most a finite number of the subscripts are negative integers. The product of two elements of H is defined to be an element of the same form containing those generators which appear in *one but not in both* of the given elements; if the given elements are the same then the product is to be the identity. This rule is obviously consistent with 1) and 2) and makes H into an abelian group every element of which is of order two.

Now if $x \varepsilon H$ is the element $x_{n_1} x_{n_2} \ldots$, then 5) $z^{-1} x z = x'$ where the generators appearing in the expansion of x' are numbered $n_1 - 1$, $n_2 - 1$, $\ldots$; this is consistent with 3). Finally, every element of G can be written as $z^m x$, where x is in H (possibly the identity) and m is an integer (possibly zero: $z^0 = e$). We have concluded our definition of the elements of the group G; products are defined using 5) to collect powers of z.

To topologize G choose neighborhoods U_m of e, $m = 1, 2, \ldots$, such that $U_m \subset H$ and contains the products of those generators x_n only for which $n > m$. The sets U_m are easily seen to be compact open subgroups of G.

Since every open subgroup must contain some elements x_n for large positive n at least, it follows from 4) that an open *invariant* subgroup must contain *all* elements x_n. Such a subgroup cannot be compact.

2.6. Approximation by separable metric groups

THEOREM. *Suppose that G is a locally compact group with identity component G_0 such that G/G_0 is compact. If neighborhoods of e, W_1, W_2, ..., are given, then there exists a compact invariant subgroup $H \subset \cap_n W_n$; G/H is separable metric and locally compact.*

Let V_0 be a compact neighborhood of e such that $G = V_0 G_0$ (2.3.1). By successive applications of 2.4, and also 1.15, we can construct a sequence of neighborhoods of e, V_1, V_2, ..., such that:

1) each V_n is compact and symmetric
2) $V_n^2 \subset V_{n-1} \cap W_n$
3) $g^{-1} V_n g \subset V_{n-1}$ for every g in V_0.

Let $H = \cap_n V_n$. Then H is a compact subgroup and $g^{-1} H g \subset H$ for every $g \varepsilon \cup_n V_0^n$. But V_0, according to 2.3.1, generates G. Therefore $g^{-1} H g \subset H$ for all $g \varepsilon G$, and H is invariant.

Let T^* denote the natural map

$$G \to G/H.$$

To show that G/H is metrizable it is enough to show that $T^*(V_n)$, $n = 1, 2, \ldots$, is a basis for neighborhoods of $T^*(e)$. Each open set in G/H containing $T^*(e)$ is the image of an open set in G containing H. Let V be an arbitrary open set in G, with $H \subset V$. We must prove that for some integer m, $V_m \subset V$.

Let $F = G - V$; then F is closed and also $H \cap F = \Phi$. Therefore

$$\Phi = (\cap_n V_n) \cap F = \cap_n (V_n \cap F),$$

and the monotonic sequence of compact sets $V_n \cap F$ has an empty intersection. It follows (1.7.1) that for some integer m, $V_m \cap F$ is empty. This concludes the argument that G/H is metrizable.

Since $G = \cup_n V_0^n$ (2.3.1) and each summand is compact (2.1) it follows that G/H is the union of a countable collection of compact metric spaces. Therefore G/H is the union of a countable collection of separable spaces (1.9) and G/H is separable. This completes the proof. The theorem is a slight but useful extension of a theorem of Kakutani and Kodaira [1].

2.6.1. LEMMA. *Suppose that G' is a locally compact group and that G is an open subgroup such that G/G_0 is compact. Suppose also that the coset-space G'/G is a countable collection of cosets. Then, given any neighborhood V of e there can be found a countable set of elements $g_n \varepsilon G'$ such that $G' = \cup_n g_n V$.*

The proof of this Lemma depends on 2.3 only. However, it will then follow from 2.6 that G' has an arbitrarily small subgroup H (invariant in G but not necessarily invariant in G') such that G'/H is a metrizable, separable coset space.

To prove the Lemma, let V_0 be a neighborhood of e in G such that $G = V_0 G_0$, and $G = \cup_n V_0^n$ (2.3.1). For each n, V_0^n is compact and may be covered by a finite number of left-translation sets of the form: gV, $g \varepsilon G$. Then G is the union of a *countable* collection of such sets. Since G' is the union of a countable collection of left-translations of G, the Lemma is proved.

2.7. Projective limits

The group G of Theorem 2.6 is a projective limit of separable metric groups. Although we shall not prove this fact we shall use it below to illustrate the concept of projective limit now to be defined.

DEFINITION. Let $A = \{a, b, c, \ldots\}$ be a directed set of indices directed by $<$; thus if $a < b$ and $b < c$, it follows that $a < c$ and to each pair a, b there is an index c, $a < c$, $b < c$. Let G_a, $a \epsilon A$, be a collection of topological groups such that if $a < b$ there is given a continuous open homomorphism T_a^b of G_b into G_a; and if $b < c$ then $T_a^c = T_a^b T_b^c$. Such a system of groups and maps is called an *inverse mapping system*. Let $\{g_a\}$ be a point of the product space PROD G_a which has the property that for $a < b$, $T_a^b(g_b) = g_a$. Such points form a group and a closed subspace of the product. This topological group is called the *projective limit* of the system.

Returning to the hypothesis and notation of 2.6, we saw that there is a sequence $V_n(e)$ of neighborhoods in G such that $T^*(V_n H)$ is a basis for open sets at $T^*(e) = eH \epsilon G/H$. Let H' be another group such that G/H' is separable metric and let $V_n'(e)$ be such that $V_n'H'$ form a basis for open sets containing H'. Then if we form

$$W_n'' = V_n \cap V_n',$$

we can determine a group H'' such that G/H'' is separable metric and $H'' \subset H' \cap H$. This observation is the basis for the final result.

We attach indices to the subgroups H such that G/H is separable metric, denoting G/H_a by G_a and consider that $a < b$ if $H_b \subset H_a$. Then maps T_a^b of G_b onto G_a exist in a natural way. The group G is the projective limit of the inverse system $\{G_a, T_a^b\}$ but we omit further details.

2.8. Real continuous functions on groups

THEOREM. *Let G be locally compact, let G/G_0 be compact and let f be a real continuous function defined on G. There exists a compact invariant subgroup H such that f is constant on cosets of H and G/H is separable metric.*

Let $V(e)$ be compact and such that $G = \cup_n V^n$. Next let R denote the space of real numbers and let O_n, $n \in I$, be a basis for open sets in R. Suppose that for some i, j, $k \in I$ the sets

$$A = f^{-1}(O_i) \cap V^k,$$
$$B = f^{-1}(O_j) \cap V^k,$$

are not empty and suppose further their closures do not intersect. Then there exists a neighborhood of e which we may denote by $W = W_{ijk}$ such that $AW \cap BW = \Phi$.

There is a countable collection of these sets W_{ijk} and if we arrange all possible sets of this kind in a sequence their intersection contains a compact invariant subgroup H such that G/H is separable metric.

Now suppose there are elements x, y, $z \in G$ with y, $z \in H$ such that

$$|f(xy) - f(xz)| = \delta > 0.$$

Because xH is compact, we have $xH \subset V^k$ for sufficiently large

$k \, \epsilon \, I$. Let O_i be an interval containing $f(xy)$ and O_j an interval containing $f(xz)$, each of length $< \delta/3$. These determine a $W_{ijk}(e)$ containing H. But $xy = a \, \epsilon \, A$, $xz = b \, \epsilon \, B$; therefore

$$x = ay^{-1} \, \epsilon \, AH \subset AW,$$
$$x = bz^{-1} \, \epsilon \, BH \subset BW.$$

This is a contradiction which completes the proof.

COROLLARY. *In the theorem above if instead of f there is given a countable set of real continuous functions $f_1, f_2, \ldots$, then H may be chosen so that each f_i is constant on each coset of H.*

For the proof take a system of W's for each f_i and select H in all of them.

2.9. Remarks on transformation groups

Let (G, M) be a transformation group and H a subgroup of G. Then (H, M) is a transformation group.

For each $x \, \epsilon \, M$, the set $H(x) \subset M$ is called an *orbit* (orbit under H, or H-orbit). The orbits are mutually exclusive subsets of M filling up M: M is a "space" of orbits. This often becomes a space in the precise meaning of the word, when open set in the orbit space is defined to be a set whose counterpart as a point set in M is open in M. It can be seen that if M is a locally compact Hausdorff space and H is compact, the orbit-space of M under H is a locally compact Hausdorff space; we shall denote it by $D(M; H)$, sometimes by M/H; and we shall call it the *decomposition space* of M by orbits of H. There is a natural map of M onto $D(M; H)$, namely $x \to H(x)$.

If H is an invariant subgroup of G, the elements of G permute the H-orbits:

$$g(H(x)) = H(g(x)) \subset M, \quad x \, \epsilon \, M, \ g \, \epsilon \, G.$$

In this action of G on $D(M; H)$, the group H leaves all points fixed.

THEOREM. *If (G, M) is a locally compact transformation group* *of a locally compact space M and H is a compact invariant sub-group of G, then G/H is a locally compact transformation group of the decomposition space D(M; H).*

The action of G/H on D(M; H) is defined by:

$$gH(H(x)) = H(g(x)), \ gH \in G/H, \ H(x) \in D(M; H).$$

The verification of this theorem is left as an exercise.

It is an exercise in the definition and topology of coset spaces to verify the following theorem. This is to be interpreted in the light of the existence of a natural map $G/N \rightarrow G/H$.

THEOREM. *Let G be a topological group, H and N closed sub-groups, with N invariant and $N \subset H$. Then there is a natural homeomorphism between the coset spaces G/H and (G/N)/(H/N).*

2.10. Remark on non-separable groups

Suppose that G/G_0 is compact and that G acts on M (G and M locally compact). Let x_n, $n \in I$, be a collection of distinct points of M. Let U_{nm}, $n, m \in I$, $n \neq m$, be a neighborhood of x_n not containing x_m, and let $W_{nm}(e) \subset G$ be such that if $g \in W_{nm}$, then $gx_n \in U_{nm}$; in particular $gx_n \neq x_m$. The neighborhoods W_{nm} are countable, and we may apply Theorem 2.6. There exists a compact invariant subgroup $H \subset \cap W_{nm}$ such that G/H is separable, metric. By the construction of H, $H(x_n) \neq H(x_m)$, $n \neq m$, and G/H is a transformation group of $D(M; H)$ in which the images of the points x_n are distinct points.

The product of uncountably many circle groups is not separable. The group of rotations of the unit circle, taken in the discrete topology, is locally compact, totally disconnected and non-separable. The natural one-one correspondence between the elements of this group and the circle-group C_1 (1.5 or 1.16) is continuous one-way, and is not open and is not a homeomorphism. This gives an example also of a transformation group of a con-nected space M where G is transitive on M and where $G_0(= e)$ is not.

2.11. Separability properties of transformation groups

THEOREM. *Let (G', M) be a locally compact transformation group. Let G be an open subgroup of G' such that G'/G is countable and such that G/G_0' is compact ($G_0' = G_0$ being the identity component of G'). Let M be locally compact and separable. Then if G' is effective, G' is separable and metrizable.*

Because M is separable it has a countable basis $\{U_m\}$, $m \in I$, of neighborhoods which may be supposed compact. For certain pairs of indices m, $n \in I$ it will be the case that

$$U_m \subset \text{interior } U_n.$$

For the same pair of indices let W_{mn} denote the set of elements of G such that

$$g(U_m) \subset \text{interior } U_n.$$

There is (2.6) a compact invariant subgroup H, $H \subset \cap W_{mn}$ for the index pairs m, n which occur as above, such that G/H is metrizable and separable.

Let h denote an arbitrary element of H, x a point of M. Let $h(x) = y$ and suppose that $x \neq y$. Then there is a compact neighborhood $U_{\overline{m}}$, $\overline{m} \in I$, of x not containing y and there is at least one compact $U_{\overline{n}}$ of x consisting of inner points of $U_{\overline{m}}$; the fact that $H \subset W_{\overline{n}\overline{m}}$ leads to immediate contradiction. Therefore no element of H moves any point of M. Since G is assumed effective, H is e. Therefore G is separable metric. Since G' is the union of a countable number of spaces homeomorphic to G (the cosets, namely, of G'/G), G' too is separable. Therefore G' is separable metric (1.22) and the theorem is proved.

COROLLARY. *Under the assumptions of the preceding theorem, except that G'/G is not required to be countable, it follows that G' is metrizable.*

Since the groups G' and G coincide locally, G' has the first countability property and the metrizability follows from 1.22.

2.12. Homogeneous spaces

A topological transformation group G is *transitive* on a space M

if for every x, $y \epsilon M$ there is a $g \epsilon G$ such that $g(x) = y$. The space M is then called a *homogeneous* space. To verify that M is a homogeneous space it is enough to know that some one point of M can be transported to every other one. Every manifold in the sense of 1.27 is a homogeneous space. The proof of this was merely indicated in 1.27.1. We shall not elaborate on it because, in general, the group of homeomorphisms of a manifold is not locally compact.

For each point x of a homogeneous space M transitive under a group G there is a *subgroup of stability* consisting of the elements of G which leave x fixed. This is a closed group which we shall usually denote by G_x. If $g(x) = y$, then $G_y = gG_xg^{-1}$, and the stability groups are conjugate.

If H is an invariant subgroup and $H \subset G_x$ for one point x of a homogeneous space M, then $H \subset G_y$ for all y in M. Then if G is effective (1.26) $H = e$. If G is effective and transitive on M the points of M are in one-one correspondence with the left-coset space G/G_x for any x of M: y in M is paired to gG_x if and only if $g(x) = y$. The map is continuous from G/G_x to M; we discuss the continuity of the inverse map in the next section.

Suppose now that the homogeneous space M *satisfies the first countability axiom*. Then a sequence of neighborhoods in M closing down on x gives rise to a sequence of neighborhoods of e in G closing down on G_x. If G is locally compact and G/G_0 is compact, such a sequence of neighborhoods contains an invariant subgroup H such that G/H is metrizable. Therefore in this case if G is effective, $H = e$, and G is metrizable.

Homogeneous spaces although topologically more general than group manifolds form a very restricted class of spaces. (See Cartan [1], Samelson [1].) For example, among two-dimensional manifolds the plane, the infinite cylinder and the torus are group manifolds; in addition, the sphere, the projective-plane, the Klein bottle, and the Mobius strip (without its customary edge) are homogeneous spaces for Lie groups (see Mostow, and earlier references there).

When a transformation group acts on a space non-transitively, then the orbit of a point x is in one-one correspondence with the homogeneous space G/G_x. This correspondence is a homeomorphism if G is compact.

2.13. Relation of homogeneous space and coset space

In general there is a distinction between the homogeneous space M on which G is transitive and the coset space G/G_x. This is shown, for instance, by the case where G is the group of integers acting on a circumference by rotating through integral multiples of an irrational multiple of 2π. Any orbit M is then dense in the circumference. However in many important cases G/G_x and M are homeomorphic. The following Theorem was proved by Arens [1]. In this connection see also Freudenthal [1, 2, 3].

THEOREM. *Let G be a locally compact group, G' an open subgroup such that G/G' is countable, and G'/G_0 ($G_0 =$ identity component so $G_0 \subset G'$ and $G_0 = G'_0$) is compact. Let G act transitively on the locally compact Hausdorff space M. Then for any $x \in M$, G/G_x is homeomorphic to M.*

In view of the theorem of 1.26 it is sufficient to prove that the map $T_1 T$ of G onto M (T maps G onto G/G_x, T_1 maps that onto M)

$$T_1 T g = g(x)$$

is open.

Let $W \subset G$ be any compact and symmetric neighborhood of e. For some countable set $g_n \in G$, $G \subset \cup_n g_n W$. Then $M \subset \cup_n g_n W(x)$, and at least one of the summands has an inner point (1.7.3; category argument). Since each $g \in G$ is a homeomorphism of M onto itself and since $(gW)(x) = g(W(x))$ it follows that $W(x)$ must have an inner point. Let $h(x)$, $h \in W$, be an inner point of $W(x)$. Then x is an inner point of $h^{-1}W(x) \subset W^2(x)$.

Let V be any open set in G and g any element in V. Now choose W compact and symmetric so that $gW^2 \subset V$. Then $W^2(x)$ has x as an inner point and $V(x) \supset gW^2(x)$ has $g(x)$ as inner point. This concludes the proof.

2.14. Transitive groups and the identity component

The following examples of coset-spaces $G/H = M$ show among other things that no simple relation with respect to transitivity need exist between the coset space G/H and the identity component G_0, even when G/H is connected. The first example is meant to illustrate what might be called the expected behavior. The first three examples are abelian.

1) Let V_1 be the additive group of reals. Let A denote a group of order 2 $(e, a, a^2 = e)$. Let $G = A \times V_1$ be the direct product of these groups. Let H_1 be the closed subgroup of G generated by the element $(e, 1)$ and let H_2 be the closed subgroup generated by the element $(a, 1)$. We see that H_1 consists of all $(e, \pm n)$ and H_2 consists of $(e, \pm 2n) \cup (a, \pm 2n + 1)$.

Now G/H_1 is a pair of circles each "covered" by one component in G and G/H_2 is a single circle covered by both components.

2) Let V_1 denote the group of the reals, and let D_2 denote the dyadic group described in example 2) of the introduction 2.0. Let $G = D_2 \times V_1$, and let H_1 denote the subgroup of G which is generated by the element $(e, 1)$ where e denotes the identity of D_2 and 1 denotes the real number one. The group H_1 consists of the set of elements $(e, \pm n)$. Then G/H_1 is a discontinuum times a circle (each circle in G/H_1 covered by one component in G). The group H_2 is defined to be the countable closed group generated by $(g_1, 1)$ where g_1, as in 2.0, denotes the so-called generator of D_2 whose powers are everywhere dense in D_2. It is left to the reader to show that G/H_2 is a single compact connected set — a *solenoid* S (of van Dantzig [2]), and no component of G covers S. Locally at each point, S has the form of a discontinuum of arcs, that is locally S is homeomorphic to a neighborhood of a point of G — and in this sense G is a cover of S; but no component of G is a cover in any customary sense.

3) Let $Z \subset V_1$ denote the subgroup of the reals consisting of the integers $\pm n$. Let $G_1 = D_2 \times Z$. Both H_1 and H_2 as defined above belong to G_1, and we can form G_1/H_2 and G_1/H_1. It can be seen that G_1/H_2 is isomorphic to D_2 and that G_1/H_1 is isomorphic to D_2.

4) This final example suggests some of the intricacies of a transformation group even when the topology is simple.

Let Q be the non-abelian group whose carrier is topologically a plane, whose elements are the transformations

$$x' = ax + b, \quad a, \ b \text{ real}, \quad a > 0$$

of a line into itself. The composition of elements of Q is given by

2.14.1 $$(a, \ b) \cdot (a', \ b') = (aa', \ ab' + b).$$

Consider Q as a transformation group of the plane by letting Q act on itself as follows: each $v \in Q$ determines the transformation

$$T_v : u \to v^{-1}uv, \quad v \in Q.$$

The orbit of a point $u = (a, \ b) \in Q$ lies on a topological line given by $a = $ constant. All such lines are orbits with a single exception. The line $a = 1$ is the union of three orbits, two open half lines and the identity $(a = 1, \ b = 0)$ which is its own orbit. On the exceptional line, corresponding to an invariant one-parameter subgroup, Q acts like a group of positive similitudes.

The group Q has one-parameter subgroups radiating from the identity. Thus for each $a \neq 1$, and b, the one-parameter group through $(a, \ b)$ is

$$\big(t, \ b(t-1)/(a-1)\big), \quad t > 0.$$

The group $(1, \ s)$, s real, is an invariant subgroup. The group Q, in the action just defined, permutes these one-parameter subgroups.

Of course Q acts on the plane E_2 in another relationship. This is the left or right translation of Q by its own elements.

2.15. The Hilbert fifth problem

It is common in geometry and many other applications that the action of a transformation group upon a manifold is described in terms of analytic functions, that is

1) the group G itself is a Lie group and when local coordinates $(g_1, \ldots, g_r)$ are taken in it which make the group operation analytic then

2) local coordinates $(x_1, \ldots, x_n)$ may be chosen in the manifold M so that the transformations of the group $g(x) = f(g; x)$ are given in the coordinates by

2.15.1 $$x_i' = f_i(x_1, \ldots, x_n; g_1, \ldots, g_r)$$

where f_i is simultaneously analytic in (x, g).

It may be of interest to incorporate a translation of a large part of Hilbert's fifth problem. This is one of 23 problems given by Hilbert in 1900 at the International Congress of Mathematics and published in *Göttinger Nachrichten* that year. The translation follows: *"V. Lie's concept of continuous transformation-groups without the assumption of the differentiability of the functions defining the group.*

"As is known Lie set up a system of axioms for geometry in which he incorporated the idea of continuous transformation groups and showed on the basis of his theory of transformation groups that his system of axioms was adequate for the building up of geometry. However, since Lie always assumes in the foundations of his theory that the functions defining his groups can be differentiated, the question is not touched on in Lie's development whether the assumption of differentiability is in fact unavoidable in the study of these axioms of geometry or whether on the contrary it is not a consequence of the concept of a group and the other geometrical axioms. These considerations, and also certain problems related to arithmetical axioms lead us to the more general problem, *to what extent the concept of the Lie continuous transformation-group is adequate for our investigations, without the assumptions of differentiability.*

"As you know, Lie defines a finite continuous transformation-group as a system of transformations

$$x_i' = f_i(x_1, \ldots, x_n; a_1, \ldots, a_r) \quad (i = 1, \ldots, n)$$

of such a nature, that any two transformations of the system

$$x_i' = f_i(x_1, \ldots, x_n; a_1, \ldots, a_r)$$
$$x_i'' = f_i(x_1', \ldots, x_n'; b_1, \ldots, b_r)$$

carried out one after the other result in a transformation which also belongs to the system and so can be represented in the form

$$x_i'' = f_i(f_1(x, a), \ldots, f_n(x, a); \, b_1, \ldots, b_r) = f_i(x_1, \ldots, x_n; \, c_1, \ldots, c_r)$$

where the $c_1, \ldots, c_r$ are certain functions of the $a_1, \ldots, a_r$ and the $b_1, \ldots, b_r$. The group property finds its expression in a system of functional equations and of itself imposes no further restrictions on the functions $f_1, \ldots, f_n, c_1, \ldots, c_r$. Nonetheless, the way in which Lie develops these functional equations necessitates the assumption of the continuity and the differentiability of the functions defining the group.

"Now as for continuity, one would certainly insist upon this condition — simply out of regard for the geometric and arithmetic applications, where the continuity of the functions which enter into the problem appears as a consequence of the continuity axiom. On the other hand the differentiability of the functions defining the group is a condition which can be formulated among the geometric axioms only in very artificial and complicated ways, and so there arises the question whether it is not possible by the introduction of appropriate new variables and parameters to transform the group into another one where the defining functions are differentiable, or at least whether it is not possible by the imposition of certain simple assumptions to achieve a transformation to a group in which the Lie methods are applicable. This reduction to an analytic group is always possible, according to a theorem proposed by Lie and first proved by Schur, if the group is transitive and if one assumes the existence of first and of certain second derivatives of the functions defining the group.

"The investigation of this problem is also interesting, I think, for infinite groups. Above all we are led to the wide and not uninteresting field of functional equations which in the past have been studied only on the assumption of the differentiability of the pertinent functions."

The last paragraph indicates the direction in which the remainder of this Fifth Problem goes, and since it lies entirely outside the scope of this book, we shall not complete the quotation. We shall not consider the "infinite continuous groups", which are not groups in the present-day sense (see E. Cartan, Collected Works, and also papers by S. Chern, by C. Ehresmann, and by Y. Matsushima).

Let us now consider the following questions, the second and third of which are asked by Hilbert:

If a locally compact group G acts effectively on a manifold M (locally-euclidean space) then

1) is G necessarily locally euclidean,

2) if the group G is locally euclidean, is it a Lie group in some appropriate coordinates.

3) If G is a Lie group, can coordinates be chosen in G and M so that the transforming functions are analytic?

The answer to 2) which can be asked, of course, without mentioning transformation groups, is yes and was first solved by the contents of two papers one by Gleason [5], the other by the authors [13]. These papers showed further that any finite-dimensional locally compact group is a generalized Lie group (this means essentially that G_0, the identity component, has arbitrarily small normal (invariant) subgroups N such that G_0/N is a Lie group). This work was extended and simplified by Yamabe [3, 4] who used the methods of Gleason to show that any locally compact group is a generalized Lie group.

The answer to 1) is unknown except in some special cases. In particular if G is connected and acts on a three-dimensional manifold then G is Lie, Montgomery-Zippin [4]. Whether 1) is true or false the question can be asked how regular G and its action must be and how closely it must resemble the familiar groups of geometry.

The answer to 3) is no. For example a group of reals can act on E^2 by having fixed $x^2 + y^2 \leq 1$ and slowly rotating the rest of E^2. This can not be analytic since if it were the existence of an open set of fixed points would imply that all points were fixed. The answer to 3) is no even if G is compact as was first shown by Bing [1] by an example of the cyclic group of order two acting on E_3 in a way which could not be differentiable. See also

Montgomery-Zippin [14]. As mentioned, 3) is true in some cases and it remains to find out when it is true or in general whether the truth resembles the differentiable case in some way.

Some of the main problems of transformation groups occur in the differentiable and analytic cases. Here, although much has been done, very little is known in the large.

As we shall see much later a locally euclidean group of dimension r has a family of one-parameter local groups filling a neighborhood of the identity and it is possible to find r of these

$$x_1(t_1),\ x_2(t_2),\ \ldots,\ x_r(t_r),\quad t_i\ \text{real},$$

such that the group product

$$x = x_1(t_1)x_2(t_2)\ \ldots\ x_r(t_r)$$

is a one-one continuous image of the r-cube

$$|t_i| \leqq a$$

for suitably small positive real a.

The numbers t_i are called *canonical coordinates of the second kind*. The group operations are analytic in these coordinates. But this solution of 2) cannot be pushed through in so direct a fashion. It has not been possible so far to solve 2) except through analyzing the structure of the whole class of locally compact groups.

In this chapter we shall set up much of the preliminary machinery. The bulk of the proofs for the structure of locally compact groups will follow in the next chapter, and will be concluded in Chapter IV.

2.16. Subgroups of Gl_n

We denote by Gl_n the group of $n \times n$ non-singular matrices with real coordinates a_{ij}, $i, j = 1, \ldots, n$; the product $C = AB$ being defined by the usual formulas:

$$c_{ik} = \textstyle\sum_{j=1}^{n} a_{ij}b_{jk}, \quad i,\ k = 1, \ldots, n.$$

In this section we shall prove the theorem of von Neumann ([1]; 1929) that a locally compact group is a Lie group if it can be mapped into Gl_n by a one-one continuous homomorphism.

This theorem, which contains the result that the closed subgroups of Gl_n are Lie groups, is a keystone in our exposition of locally compact groups.

The proof depends upon the properties of the exponential map:

$$A \to \exp A$$

which connects the additive group M_n of $n \times n$ real matrices and the group Gl_n. This map is defined by the exponential series formula, applied to matrices. Let the matrix A be (a_{ij}), and let $(a_{ij})^m = A^m$ denote the m-th power of the matrix. Then $\exp(a_{ij})$ is defined by the series expansions:

$$\exp(a_{ij}) = (\delta_{ij}) + (a_{ij}) + (1/2!)(a_{ij})^2 + \ldots + (1/m!)(a_{ij})^m + \ldots,$$

with the usual matrix multiplication and addition, and where (δ_{ij}) is the unit $n \times n$ matrix.

Let the coordinates a_{ij} be numerically smaller than some $r > 0$. It can then be seen by an inductive argument that the components of the matrix $(a_{ij})^m$ are numerically less than $n^m r^m$. Therefore each component of the matrix $\exp A$ is defined and is less in absolute value than

$$\exp nr = 1 + nr + (1/2!)(nr)^2 + \ldots.$$

If we let A range over those matrices of M_n all of whose n^2 coordinates are numerically $\leq r$ then each of the n^2 coordinates x_{ij} of $\exp A$ is a uniformly convergent power series in the coordinates of A and is an analytic function of those coordinates.

The Jacobian of the n^2 functions x_{ij} with respect to the n^2 arguments a_{ij} is equal to one at the origin of components of M_n. To see this arrange the components a_{ij} of A in some linear order:

$$A = (u_1, \ldots, u_{n^2}) \; \varepsilon \; M_n$$

and arrange the components of the matrix $\exp A$ in the corresponding order:

$$\exp A = (v_1 \ldots, v_{n^2}) \; \varepsilon \; Gl_n.$$

Take as the coordinates of matrices of Gl_n not the actual components x_{ij} but rather the differences: $x_{ij} - \delta_{ij}$. Now, to calculate each of the n^4 terms $\partial v_i/\partial u_j$ at the origin we may let all u_i's be zero except one, say $u_j = u$. The terms in exp A of the first order in u show that $\partial v_i/\partial u_j = \delta_{ij}$, $i, j = 1, \ldots, n^2$, and the Jacobian at the identity

$$A = (0, \ldots, 0)$$

is simply the unit $n^2 \times n^2$ determinant.

It follows from the implicit function theorem that there is a compact neighborhood U_0 of the origin of M_n, defined by:

$$|u_i| \leq r' \text{ for some } r' > 0,$$

which is mapped topologically onto a neighborhood V_0 of the identity matrix in Gl_n. In what follows we stay within these neighborhoods.

The homeomorphism which has just been constructed cannot be an isomorphism for $n > 1$ since M_n is abelian and Gl_n is not. However the map carries one-parameter groups of M_n into one-parameter groups in Gl_n. This important fact is seen as follows. Let A be a matrix of M_n and let r and s be real numbers. Then the product in Gl_n of exp rA and exp sA is given by

$$\exp rA \exp sA = \left(\sum_{k=0}^{\infty} r^k A^k/k! \right)\left(\sum_{j=0}^{\infty} s^j A^j/j! \right)$$

$$= \sum_{m=0}^{\infty} \sum_{i=0}^{m} \frac{r^i s^{m-i}}{i!(m-i)!} A^m$$

$$= \sum_{m=0}^{\infty} \frac{(r+s)^m}{m!} A^m$$

$$= \exp(r+s)A.$$

In the group M_n each element A for each integer m has one and only one m-th root namely the matrix $(1/m)A$ which we may also write as A/m. Since the exponential is one-one from U_0 to V_0, it follows that *each element of V_0 has one and only one m-th root in V_0, for every integer m.* This shows that there are no other one-parameter local groups in V_0 than the ones which we get as images of the linear groups tA in M_n, t real.

There is a formula expressing exp $(A + B)$ in terms of exp A, exp B. In order to derive this formula we consider the other standard definition of the exponential.

The definition of exp A has been given as

1) $\exp A = I + A + A^2/2! + \ldots$

but it can be shown that the following definition is equivalent

2) $\exp A = \lim (I + A/m)^m$.

In order to see the equivalence we proceed as follows. From 1)

$$3) \quad \exp A = I + A + A^2/2! + \ldots + A^k/k! + R(k)$$
$$= S(k) + R(k),$$

where $R(k)$ is a matrix which tends to zero as k goes to infinity. Let the matrix $T(m) = (I + A/m)^m$; for $m > k$,

$$4) \quad T(m) = I + A + \frac{m(m-1)}{m^2} \frac{A^2}{2!} + \ldots$$
$$+ \frac{m(m-1) \ldots (m-(k-1))}{m^k} \frac{A^k}{k!} + E(m, k).$$
$$= D(m, k) + E(m, k).$$

Let $\epsilon' > 0$ be given. Then because the exponential series 1) is uniformly convergent on compact sets, it follows that there is an integer N such that for any integer q

$$5') \quad \left| \sum_{k=N}^{N+q} \frac{A'^k}{k!} \right| < \epsilon'$$

for all $A' \varepsilon U_0$. For the moment we are using the absolute-value sign to denote the maximum value of any component of the enclosed matrix. The freedom to vary A in U_0 will be useful in a later Corollary.

Hence given any $\epsilon > 0$ there is an integer N such that for all $k > N$ and all $m > k$ we have

5) $| T(m) - D(m, k) | = | E(m, k) | < \epsilon/3$.

Further let the choice of N be such that if $k > N$

6) $| \exp A - S(k) | < \epsilon/3$.

Now take a fixed $k > N$; with k fixed choose M so that if $m > M$

7) $\quad |S(k) - D(m, k)| < \epsilon/3.$

Combining 5), 6), 7), we obtain

$$|\exp A - T(m)| < \epsilon$$

for all $m > M$. This proves 2).

LEMMA. *Let A and B be matrices of M_n with exp A, exp B the associated elements of $Gl(n, R)$. Then*

$$\exp (A + B) = \lim (\exp A/m \exp B/m)^m.$$

Let

$z(m) = \exp A/m \exp B/m$

$= (I + (1/m)A + (1/2!m^2)A^2 + \ldots)(I + (1/m)B + (1/2!m^2)B^2 + \ldots)$

$= (I + (1/m)A + (1/m^2)A')(I + (1/m)B + (1/m^2)B')$

$= I + (1/m)(A + B) + (1/m^2)C,$

where A', B', C are matrices depending on m but uniformly bounded; in fact if r' is a bound for the elements of A and B then there is a bound α for all elements of A', B', and C depending only on r' and on the order of A and B. Hence for $m > k$

8) $\quad z(m)^m = I + (A + B + (1/m)C) + \dfrac{m(m-1)}{m^2} \dfrac{(A+B+(1/m)C)^2}{2!} + \ldots$

$$+ \dfrac{m(m-1)\ldots(m-(k-1))}{m^k} \dfrac{(A+B+(1/m)C)^k}{k!} + R(m,k)$$

$$= Q(m, k) + R(m, k).$$

We know

$$\exp (A + B) = \lim (I + (1/m)(A + B))^m,$$

and we let

$$T(m) = (I + (1/m)(A + B))^m.$$

For $m > k$ we have

9) $T(m) = I + (A + B) + \dfrac{m(m-1)}{m^2} \dfrac{(A+B)^2}{2!} + \ldots$

$\quad + \dfrac{m(m-1) \ldots (m-(k-1))}{m^k} \dfrac{(A+B)^k}{k!} + E(m, k)$

$\quad = D(m, k) + E(m, k).$

Given $\epsilon > 0$, there is an integer K such that for all m if $m > k > K$,

9') $\quad | z(m)^m - Q(m, k) | < \epsilon/4,$

$\quad | D(m, k) - T(m) | < \epsilon/4.$

It is worth noting that 9') holds for all A, $B \; \epsilon \; U_0$ if the integer K is large enough (in view of the remarks accompanying 5')). Now fix a value for k and then choose an integer N' so that

$$| Q(m, k) - D(m, k) | < \epsilon/4$$
$$| T(m) - \exp (A + B) | < \epsilon/4,$$

for $m > N'$. Now for $m > N' > k$

$$| z(m)^m - \exp (A + B) | < \epsilon$$

and this completes the proof of the Lemma.

COROLLARY. *Suppose that* A, A_m, B, $B_m \; \epsilon \; M_n$, $m \; \epsilon \; I$ *and that*

$$\lim A_m = A, \; \lim B_m = B.$$

Then:

$$\lim (\exp A_m/m \; \exp B_m/m)^m = \exp (A + B).$$

The proof is contained in the proof of the Lemma, in view of the uniform convergence of the exponential (see remarks accompanying 5'). The details are left to the reader.

LEMMA. *Let* $M_n = L' + L''$ *be a direct decomposition of* M_n *into complementary linear subspaces of dimensions* n' *and* n'' *respectively* $(n' + n'' = n^2)$. *Then every element* z *of* Gl_n *near enough to the identity is a unique product of the form:*

10) $\quad z = \exp A' \exp A''$

with A' *in* $U_0 \cap L'$ *and* A'' *in* $U_0 \cap L''$.

Let A_k, $k = 1, \ldots, n'$, be a basis for elements of L' and let A_k, $k = n' + 1, \ldots, n^2$, be a basis for elements in L''. Then every matrix A in M_n is a unique real linear form:

11) $\quad A = \Sigma_{k=1}^{n'} t_k A_k + \Sigma_{k=n'+1}^{n^2} t_k A_k.$

Let e_1 denote the identity of Gl_n. Let V_1 be a neighborhood of e_1 in Gl_n whose square is in V_0 and let U_1 be the counter-image set in M_n under the exponential map as restricted to U_0. Now if A' is in $U_1 \cap L'$ and A'' is in $U_1 \cap L''$, then $z = \exp A' \exp A''$ belongs to V_0 and is of the form $\exp A$ for some A in U_0. Then z *may be given two sets of coordinates*, the numbers $v_1, \ldots, v_{n^2}$ defined earlier as coordinates of $\exp A$ and the new numbers $t_1, \ldots, t_{n^2}$ all of which are zero at $e_1 \, \epsilon \, Gl_n$.

To find the Jacobian of the transformation at e_1 we calculate the n^4 partial derivatives: $\partial v_i / \partial t_j$. For fixed j let all $t_i = 0$ except $t_j = t$. Now whatever value j may have z is of the very special form $\exp t A_j$. The explicit form of the exponential shows that the derivative of this matrix with respect to t is the matrix A_j. The desired derivative $\partial v_i / \partial t_j$ is simply the v_i-th coordinate of the particular matrix A_j: we denote this by $v_i^{(j)}$, $i, j = 1, \ldots, n^2$.

The $n^2 \times n^2$ Jacobian of the transformation of coordinates in Gl_n at the origin e_1 coincides with the determinant of the system of equations indicated in 11). If u_i, $i = 1, \ldots, n^2$, are the coordinates of A, then 11) takes the form of n^2 equations in n^2 unknowns: t_k.

12) $\quad u_i = \Sigma_{k=1}^{n^2} t_k v_i^{(k)}, \quad i = 1, \ldots, n^2.$

The determinant of this system cannot be zero because, by construction, the set of matrices A_k is a basis for elements of M_n.

Since the Jacobian does not vanish at e_1 in Gl_n

$$\frac{D(v_i)}{D(t_j)} \neq 0, \quad t_1 = t_2 = \ldots = t_{n^2} = 0$$

It follows by the implicit function theorem that $z = \exp A' \exp A''$ ranges in a one-one way over some neighborhood of e_1 in Gl_n. This proves the Lemma.

2.16.1. Having analyzed the exponential map in some detail we turn now to those subgroups of Gl_n which can be the one-one homomorphic images of locally compact groups. The closed subgroups of Gl_n constitute the most natural examples of such image-groups. A different kind of example is got by mapping the group of the integers so that the image is everywhere dense in a circle group. A similar, less trivial example is the type of one-parameter global group on a torus which winds round and round without closing to a circle.

LEMMA. *Let G be a locally compact group and let φ be a one-one continuous homomorphic map of G into Gl_n. Then there is a linear subspace L' in M_n and a compact neighborhood $V' \subset G$ of the identity in G such that the set $\varphi(V')$ coincides in a neighborhood of e_1 with the exponential map of $L' \cap U_0$ (U_0 as on p. 73).*

Since $\exp(U_0)$ is a neighborhood of e_1 and φ is continuous, there is a compact symmetric neighborhood $V \subset G$ of the identity of G such that

$$\varphi(V) \subset \exp(U_0).$$

Since V is compact and φ is one-one and continuous, it follows that φ is a homeomorphism on V. If V is discrete (this would be the case in the trivial example mentioned above) then $\varphi(V)$ is also discrete and there is nothing to prove since the zero of M_n can be regarded as a (zero-dimensional) linear subspace. Accordingly we suppose that $\varphi(V)$ contains a sequence of distinct elements $x_1, x_2, \ldots$ converging to e_1, the identity of Gl_n. Each x_n belongs to a one-parameter local group $X_n(t)$ which is contained in $\exp(U_0)$ and meets the boundary of $\exp(U_0)$. It follows from this that there is a sequence of smallest integers, $k_1, k_2, \ldots$, such that

$$x_n^{k_n} \, \varepsilon \, \varphi(V) \text{ but } x_n^{k_n+1} \, \xcancel{\varepsilon} \, \varphi(V), \ n = 1, 2, \ldots$$

We may suppose, without loss of generality, that the sequence $x_1^{k_1}, x_2^{k_2}, \ldots$ converges to an element of x^* of $\varphi(V)$. Since the sequence $x_1^{k_1+1}, x_2^{k_2+1}, \ldots$ also converges to x^* and since the identity of G is an inner point of V, it follows that x^* is not e_1. Then, *first*, x^* determines a unique one-parameter local group $X(t)$ which we may suppose so "normalized" that $x^* = X(1)$. *Second*, the integers $k_1, k_2, \ldots$ increase without limit. We suppose the parameter on each group $X_n(t)$ normalized so that $X_n(1) = x_n^{k_n}$, and then $x_n = X_n(1/k_n)$. By looking at the counter-images in M_n one sees easily that, for any choice of t' in $[-1, 1]$, $X(t')$ is a limit of a sequence of elements $x_n^{i_n}$ where $i_n \leq k_n$ and $\lim i_n/k_n = t'$. This shows that $X(t)$ belongs to $\varphi(V)$ for $|t| \leq 1$.

Now let us denote by L' the set of elements A of M_n such that some local subgroup of $\exp tA$ is mapped into $\varphi(V)$. Then if $A' \varepsilon L'$, $\exp tA' \varepsilon \varphi(G)$ for all t and there is a maximum $t' > 0$ such that $\exp tA' \varepsilon \varphi(V)$ for $|t| \leq t'$. The element $\exp t'A'$ belongs to $\varphi(Bdry\ V)$. We have proved above that L' contains a non-trivial element. We show next that L' is a linear subspace of M_n. It is clear that if $A \varepsilon L'$ then $rA \varepsilon L'$ for every real r.

Let $A, B \varepsilon L' \cap U_0, A + B \varepsilon U_0$. Then, by the definition of L', $\varphi(V)$ contains every element of $\exp tA$ and $\exp tB$, $|t| \leq c$ for some $c > 0$. Then for all large n the

$$x_n = \exp A/n \exp B/n$$

belong to $\varphi(V)$; clearly, the sequence of these elements converges to e_1. If x_n were equal to e_1 (for infinitely many values of n) we would conclude from $\exp (A + B) = \lim (\exp A/n \exp B/n)^n$ that $A + B$ is the *zero* and certainly belongs to L'. We suppose now that the x_n are different from e_1 and then it follows by an earlier argument that there is a sequence of integers $k_1, k_2, \ldots$ such that some subsequence of

$$x_1^{k_1}, x_2^{k_2}, \ldots$$

converges to $x^* \varepsilon \varphi(V)$, $x^* \neq e_1$; and x^* belongs to a uniquely determined one-parameter local group $X(t) \subset \varphi(V)$.

If $k_n \geq n$ for infinitely many values of n then for those values $(\exp A/n \exp B/n)^n$ belongs to $\varphi(V)$. Therefore in this case $\varphi(V)$ contains $\exp (A + B)$, and $\exp t(A + B)$ coincides with $X(t)$ suitably normalized. Then $A + B \varepsilon L'$. If $k_n \leq n$, then for some sequence $I'' \subset I$, $\lim k_n/n = k \leq 1$ exists as n ranges over I''. Set

$A_n = k_n A/n$, $B_n = k_n B/n$. Then

$$\lim (\exp A/n \exp B/n)^{k_n}$$
$$= \lim (\exp A_n/k_n \exp B_n/k_n)^{k_n}$$
$$= \exp k(A + B), \text{ by corollary p. 76.}$$

Now it is clear that $\varphi(V)$ contains $\exp kt(A + B)$ for $|t| \leqq 1$ and therefore $A + B \,\varepsilon\, L'$. Then we have shown that L' is a linear space.

Let L'' denote a linear subspace of M_n complementary to L' and choose a basis for M_n: $A_1, \ldots, A_{n'}, A_{n'+1}, \ldots, A_{n^2}$ as in 11) above. Let $U^* \subset U_0$ denote the set of elements satisfying

13) $A = \Sigma_1^{n^2} t_k A_k$ $|t_i| \leqq c$,

for $c > 0$ chosen as follows. *First*, in accordance with the preceding Lemma, choose c so that if $A \,\varepsilon\, U^*$ then $\exp A = \exp A' \exp A''$ as in 10). *Second*, since the *boundary* of V (this set is in G) is compact and does not contain the identity, then $\varphi(\text{Bdry } V) \subset Gl_n$ is closed and does not contain e_1; therefore we can require c to be so small that $\exp A \,\notin\, \varphi(\text{Bdry } V)$, for $A \,\varepsilon\, U^*$. In this case it is also true that $\exp tA \,\notin\, \varphi(\text{Bdry } V)$ for $|t| \leqq 1$.

Let $L_1 \subset Gl_n$ denote the image of $U^* \cap L'$ under the exponential map. It follows at once from the choice of c that $L_1 \subset \varphi(V)$. We need to prove that $\varphi^{-1}(L_1) \subset G$ is a neighborhood of the identity. This is equivalent to showing that if $\exp A_1$, $\exp A_2$, ... is a sequence of elements of $\varphi(V)$ converging to e_1, then $\exp A_n \,\varepsilon\, L_1$ for sufficiently large n.

The sequence $A_1, A_2, \ldots$ converges to the *zero* of M_n and for large n, $A_n \,\varepsilon\, U^*$ and $\exp A_n = \exp A_n' \exp A_n''$ as in 10) for suitable A_n', A_n''. For any convergent *subsequence* of $A_1', A_2', \ldots$, the corresponding *subsequence* of $A_1'', A_2'', \ldots$, converges and then both subsequences converge to $e_0 \,\varepsilon\, M_n$; this is the only element common to L' and L''. Therefore both *sequences* converge to e_0. Since $\exp A_n$ and $\exp A_n'$ are in $\varphi(V)$ and are near e_1 for large n, it follows that $\exp A_n''$ is also in $\varphi(V)$ for large n.

Set $x_n = \exp A_n''$, for all n. Now precisely as in an earlier argument, there is a subsequence of some sequence of powers:

$x_1^{k_1}$, $x_2^{k_2}$, ..., all belonging to $\varphi(V)$, which defines a non-trivial one-parameter local group $X(t) \subset \varphi(V)$. Now, however, $X(t)$ is the image of elements in L'' and this is a contradiction. This concludes the proof that $\varphi^{-1}(L_1)$ is the neighborhood V' required by the Lemma.

It is interesting that this neighborhood is filled by a family of one-parameter local groups intersecting only at e, and running to the boundary. The existence of such a neighborhood, as we shall see much later, characterizes Lie groups.

2.16.2.   Finally we show that the group G described in the Lemma above is a Lie group, $\varphi^{-1}(L_1)$ being a suitably coordinatized neighborhood. Let $\exp A$, $\exp B$, $\exp C \, \varepsilon \, L_1$ be such that

$$\exp C = \exp A \, \exp B.$$

We must show that the coordinates of $\exp C$ are analytic functions of the coordinates of $\exp A$ and $\exp B$. Let r_k, s_k, t_k, $k = 1, \ldots, n^2$, be the coordinates of the matrices A, B, C respectively, as expressed by 11). The last $n^2 - n'$ of these coordinates are zero.

Now r_k, s_k, t_k, $k = 1, \ldots, n'$ are the coordinates in L_1 of $\exp A$, $\exp B$, $\exp C$, by definition. We must trace the dependence of each t_k upon the set of r_k, s_k, $k = 1, \ldots, n'$. The n^2 ordinary components of the matrices A and B are linear functions with constant coefficients of the coordinates r_k and s_k, respectively. The components of $\exp A$, $\exp B$ are analytic functions of the components of A, B and therefore analytic functions of the coordinates r_k, s_k. The components of $\exp C$ are bilinear functions of the components of $\exp A$ and $\exp B$, and therefore analytic in the r_k and s_k jointly. The n^2 components of the matrix C are analytic functions of the components of $\exp C$: this follows from the fact that the components of $\exp C$ are n^2 analytic functions of those of C with non-vanishing Jacobian in a neighborhood of the origin of Gl_n. Finally, the coordinates of the matrix C are linear functions of the components of C, as determined by the unique solutions of 11): actually, the n^2 components of C determine n^2 numbers t_k, but the last $n^2 - n'$ are zero since $\exp C$ is in L_1 by hypothesis.

This completes the proof of the principal theorem of this section, as follows.

THEOREM: *A locally compact group G is a Lie group if it can be mapped into Gl_n by a continuous one-one homomorphism φ.*

The subgroup $\varphi(G) \subset Gl_n$ is not necessarily closed unless G is compact. However, every closed subgroup of Gl_n satisfies the hypotheses of the theorem and is a Lie group. In this way one sees that the *orthogonal group* and other familiar groups are Lie groups. The exponential map is frequently useful for showing more directly that a particular group is a Lie group. By way of example, let us look at the orthogonal group in n real variables.

A real matrix $B = (b_{ij})$ is called *orthogonal* if the *transpose* matrix is also the *inverse*. The set of such matrices is a compact subgroup of Gl_n. Let us denote it by R_n.

Recall that for any matrix A, exp tA is a one-parameter subgroup of Gl_n and therefore:

$$\exp(-A) = (\exp A)^{-1}.$$

Furthermore, for all matrices, directly from the definitions:

$$\exp(\text{transpose } A) = \text{transpose } (\exp A).$$

Now let $A = (a_{ij})$ be a skew-symmetric matrix:

$$a_{ij} = -a_{ji}, \quad i, j = 1, 2, \ldots, n.$$

Then exp A is orthogonal, because if A is skew-symmetric:

$$\exp(\text{transpose } A) = \exp(-A)$$

and, therefore, combining the relations above, we get

$$\text{transpose } (\exp A) = (\exp A)^{-1}.$$

One can see, as follows, that all orthogonal matrices near the unit-matrix are images of skew-symmetric matrices. Since R_n is transitive on the unit-sphere in n-space and the group of stability is R_{n-1}, it follows inductively that the dimension of R_n is $(n^2 - n)/2$. This is also the dimension of the *additive group* of skew-symmetric matrices and it follows that R_n is (locally) the one-one image of that group under the exponential map.

2.16.3. If $K = \varphi(G)$, G as in the preceding theorem, is *abelian* then the exponential map establishes an isomorphism between a neighborhood of the linear vector space $L' \subset M_n$ and a neighborhood of e in K. *For* if A and B are in M_n and $\exp tA$ and $\exp tB$ commute for all t, then

$$\exp(A + B) = \lim (\exp A/m \exp B/m)^m = \exp A \exp B.$$

If K is abelian and connected then the next theorem will show that it has the structure of a product (1.11.1) of a vector group and a toral group.

THEOREM. *If K is a connected locally compact abelian group which is locally isomorphic to V_n, then*

$$K = T_k \times V_r, \quad k + r = n$$

with T_k a toral group of dimension k.

Let f denote an isomorphic map of a neighborhood Q of the zero of V_n onto a neighborhood U of e in K. Extend f to a homomorphic map of V_n onto K as follows. For each $z \varepsilon V_n$ there is a sufficiently large integer m' such that for $m \geq m'$ the m-th root of z (call it z_m, and then $mz_m = z$) belongs to Q. Now let $f(z)$ be defined as $mf(z_m) \varepsilon K$. It follows readily that $f(z)$ does not depend on the choice of $m \geq m'$.

Let $P \subset V_n$ be a neighborhood of the zero such that $P^2 \subset Q$. Now if x, y, z are elements of V_n such that $z = x + y$ and if they have m-th roots x_m, y_m, z_m in P then

$$\begin{aligned}
f(z) = mf(z_m) &= mf(x_m + y_m) \\
&= m[f(x_m) + f(y_m)] \\
&= mf(x_m) + mf(y_m) = x + y.
\end{aligned}$$

This shows that the extension of f is a homomorphism. Since $f(Q) = U$ is a homeomorphism it follows that f is continuous and open, and K is isomorphic to V_n/D where D is the kernel of the map. We see that D is discrete because $D \cap Q$ contains only the zero of V_n.

Now what remains to be proved is equivalent to the well-known fact that if D is a discrete subgroup of V_n, then V_n/D is the product of a toral group and a vector group. Let V_k denote the linear subspace of V_n of smallest dimension $k \leq n$ which contains D. There is a direct decomposition of V_n:

$$V_n = V_k \times V_r.$$

Since $D \subset V_k$, and $V_k \cap V_r$ is the identity, it follows that V_n/D is a direct product, and

$$K = f(V_k) \times f(V_r).$$

Now $f(V_r)$ is a space isomorphic to V_r. We must show that $f(V_k) = V_k/D$ is a toral group.

If $k = 0$ there is nothing to prove; suppose $k > 0$. Since V_k is the smallest linear subspace of V_n containing D we can find k one-dimensional linearly independent vector groups $X_1, \ldots, X_k$ such that $X_i \cap D$, $i = 1, \ldots, k$, contains at least two elements. Therefore $X_i \cap D$ is isomorphic to the group of integers. Let x_i denote a generating element of $X_i \cap D$. Every element x of V_k can be written in one and only one way as

1) $\quad x = r_1 x_1 + \ldots + r_k x_k$, r's reals.

Since $x_i \, \varepsilon \, D$, every element x' such that:

2) $\quad x' = m_1 x_1 + \ldots + m_k x_k$, m's integers

is an element of D. If it were true conversely that every $x' \, \varepsilon \, D$ could be written as in 2) then it would be clear that V_k/D is a product of k circle groups and the proof of the theorem would be completed. But, in general, we need to construct a new basis $y_1, \ldots, y_k$ which will have that property.

For each $i = 1, \ldots, k$ choose an element y_i of D as follows. There is a bounded set $S_i \subset V_k$:

3) $\quad S_i = \{y; \ y = \Sigma_{j=1}^{i} r_j x_j, \ 0 \leq r_j \leq 1, \ r_i > 0\}$.

Since D is discrete, $D \cap S_i$ is finite; and it is not empty because $x_i \, \varepsilon \, S_i$. Let y_i be one of the elements of $D \cap S_i$ *for which the coefficient r_i is as small as possible.*

It is clear that the set of elements $y_1, \ldots, y_k$ forms a basis for V_k (using real coefficients); it must be proved that they generate D (using integer coefficients). Take $y \, \varepsilon \, D$; by adding to y suitable (integral) multiples of the elements y_i we get an element $y' \, \varepsilon \, D$ for which

$$y' = \Sigma_1^k r_j y_j, \ 0 \leqq r_j < 1, \ j = 1, \ldots, k.$$

If all $r_j = 0$, then y is in the group generated by $y_1, \ldots y_k$, as asserted. Suppose that for some $i \leqq k$, $r_i > 0$ but $r_j = 0$ for $j > i$. From $y' = r_1 y_1 + \ldots + r_i y_i$ and the fact that $0 < r_i < 1$, it is clear that the expansion of y' in terms of $x_1 \ldots x_i$ has a *smaller* last coordinate than the expansion of y_i (in terms of the $x_1 \ldots x_i$); this leads to a contradiction to the choice of y_i and completes the proof.

2.17. Invariant integral

In 1933 Haar [1] proved that every locally compact separable group possessed a measure function which was invariant under left translation, and from this a Lebesgue integral can be defined. In Lie groups the existence of the integral was known earlier and had been used by Peter and Weyl [1] to analyze the representations of a compact Lie group. Von Neumann [4] showed that a left invariant measure is unique up to multiplication by a positive real number.

The proof below for the existence of an invariant integral follows Haar's proof in the form given to it by A. Weil [1]. This uses the Tychonoff Theorem and makes no distinction between the separable and general cases. We give the definition in this way but for developing the details we consider first the separable case. We then use this case and the fact that any locally compact group has many separable factor groups to extend to the general case. We do this because the separable case is adequate for our purposes and because it may be of interest that this procedure gives the integral in the non-separable case without the use of the general Tychonoff theorem.

We work with the larger class of local groups; we develop only the propeities of the integral needed for our applications.

Given any group G with product ab we may define a new group G^* with the same elements and topology and a new multi-plication $a \times b$ defined as follows

$$a \times b = ba.$$

The group G^* has the same topological properties as before.   If G^* has a left invariant integral then G has a right invariant integral, so that the existence of a right invariant integral can be deduced immediately from the existence of a left invariant integral.

2.17.1.   A carrier of a function is any set outside of which the function vanishes; occasionally we use *the* carrier to mean the closure of the set on which f does not vanish.

Let G be a locally compact group (or local group) and let W be a compact symmetric neighborhood of e (such that W^6 is defined if G is a local group).   We shall continue to put in parentheses most of the remarks which apply mainly to local groups.

Let C denote the class of continuous real functions on G having compact carriers (carriers interior to W), and let C be topologized by the metric

$$d(f, g) = \text{l.u.b.} \mid f(x) - g(x) \mid, \text{ all } x \, \varepsilon \, G; \; f, \; g \, \varepsilon \, C.$$

Let C^+ denote the non-negative functions in C.   For f and g in C^+ there may be an integer n and elements $a_1, \ldots, a_n$ in G (in W^2) and non-negative numbers $r_1, \ldots, r_n$ such that for x (in W) and hence for all x

$$f(x) \leqq \Sigma r_k g(a_k^{-1} x).$$

If so, $(f : g)$ is defined to be the greatest lower bound of all Σr_k for all such sets of elements and numbers, and if not, $(f : g)$ is defined to be plus infinity.   For f, $g \, \epsilon \, C^+$, one sees that

$$(f : g) \geqq (\text{l.u.b. } f)/(\text{l.u.b. } g)$$

and $(f : g)$ is finite when g is not identically zero.   It is greater than zero if f also is not identically zero. If f_1 and g are not zero,

$$(f : g) \leqq (f : f_1)(f_1 : g)$$

and

$$\frac{1}{(f_1 : f)} \leqq \frac{(f : g)}{(f_1 : g)} \leqq (f : f_1).$$

2.17.2.   Let f_1 be a fixed function in C^+ not identically zero.

To each function f in C^+ we now associate the closed interval I_f on the real axis defined by

$$I_f : \frac{1}{(f_1 : f)} \leq x \leq (f_1 : f).$$

Using the Tychonoff Theorem (1.8.1.) we form the compact set P

$$P = \text{PROD}_f \, I_f.$$

Next, let $V \subset W$ be a neighborhood of e. Then let g not identically zero be an element of C^+ with carrier in V and consider the ratio:

$$\frac{(f : g)}{(f_1 : g)}.$$

It follows from the inequality preceding this section that the ratio $(f : g)/(f_1 : g)$ belongs to I_f for every g chosen as above. Thus g determines a unique point in P whose coordinates are those ratios. Then the neighborhood $V \subset W$ determines the following closed subset F_V of P, namely the closure of the set of all points of P associated with functions g as just described.

Finally, because P is compact there is at least one point of P in $\cap F_V$: the intersection of all sets F_V for all neighborhoods V of e, subject to $V \subset W$. To each choice of a point in $\cap F_V$ we get an *integral*. Thus let $\{r_f\}$ denote a particular choice of a point of P in $\cap F_V$, the f-coordinate of this point being the real number r_f. Then we shall define $\int f$ by

$$\int f = r_f.$$

We shall sometimes write $\int f(x)dx$ for $\int f$.

However instead of developing the properties of the integral thus defined we now return to take up the separable case, and carry out the procedure mentioned earlier. To continue from the above definition would use many similar details.

2.17.3. Assume now that G is *separable metric* and let W be a neighborhood as above. Let $\{W_n\}$ be a basis for open sets including e (assume $W_n \subset W$). Choose functions g_n such that

a) $0 \leqq g_n \leqq 1$
b) $g_n(e) = 1$
c) g_n vanishes outside W_n.

It has been shown that the set of real continuous functions defined on a compact metric space is separable. Hence C^+ is seen at once to be separable if G is a local group. However C^+ is also separable in case the separable metric G is a global group, for in this case G is seen to be the union of a countable increasing family of compact sets so that C^+ is the union of a countable collection of separable subsets, and this implies that C^+ is separable.

Hence there is a countable family $\{f_n\}$ which is a dense subset of C^+. We shall suppose that one of these, f_1, is identically 1 on a neighborhood of e. It is convenient to have $\{f_n\}$ closed under addition of its elements and also under multiplication by a positive *rational* number and we choose $\{f_n\}$ in this way, adjoining new functions if necessary.

Let $I^* \subset I$ be any infinite sequence such that

$$\lim_{i \in I^*} \frac{(f_n : g_i)}{(f_1 : g_i)}$$

exists when i runs through I^* for all functions f_n, $n \in I$. By the diagonal process such a sequence I^* can be seen to exist. This limit for each n is denoted by $\int f_n(x)dx = \int f_n$.

Whenever

A) $$\lim_{i \in I^*} \frac{(f : g_i)}{(f_1 : g_i)}$$

exists for a function f in C^+ it is denoted by $\int f$. By a preceding formula

1) $$1/(f_1 : f) \leqq \int f \leqq (f : f_1)$$

so that in particular the integral is always finite.

If t is a non-negative real constant then $(tf : g) = t(f : g)$. Therefore if $\int f$ exists then $\int tf$ exists and is equal to $t\int f$. It follows that $\int tf_1 = t$.

From $(f + f' : g) \leqq (f : g) + (f' : g)$ it follows that

$$\textstyle\int f + \int f' \geqq \int (f + f')$$

if all three exist. If in addition $f(x) \leqq f'(x)$, $x \, \epsilon \, G$, then $\int f \leqq \int f'$.

Suppose the integral exists for f, f', F_1, and that $f \leqq f' + tF_1$. Then $(f : g_i) \leqq (f' : g_i) + t(F_1 : g_i)$ and since limits exist,

$$\textstyle\int f \leqq \int f' + t\int F_1.$$

Now if $|f - f'| \leqq tF_1$, it follows that $f' \leqq f + tF_1$. Hence

$$\textstyle\int f' - t\int F_1 \leqq \int f \leqq \int f' + t\int F_1$$
$$\textstyle|\int f - \int f'| \leqq t\int F_1.$$

Hence if the integral exists for each element of a uniformly converging sequence $h_n \, \epsilon \, C^+$, with carrier h_n in a fixed compact set K, and exists also for the limit function $h \, \epsilon \, C^+$, then

$$\textstyle\lim \int h_n = \int \lim h_n = \int h.$$

To see this choose F_1 in $\{f_n\}$ with a positive lower bound on K.

Let h be any function in C^+. Then h is the uniform limit of functions h_n in the set $\{f_n\}$, carrier h_n in a fixed compact set, so that $\int h_n$ exists for all n. We shall show that $\int h$ exists. If the ratios

B) $\dfrac{(h : g_i)}{(f_1 : g_i)}$

do not converge with increasing $i \, \epsilon \, I^*$, there must exist two subsequences $I_1^* \subset I^*$ and $I_2^* \subset I^*$ such that $B)$ has a limit, call it $\int_1 h$ for $i \, \epsilon \, I_1^*$ and another limit, call it $\int_2 h$ for $i \, \epsilon \, I_2^*$. Each of these subsequences defines an integral $\int_1 f$ and $\int_2 f$ with the properties already established for $\int f$; and $\int_1 h_n = \int_2 h_n = \int h_n$ if h_n belongs to the family $\{f_n\}$. Then

$$\textstyle\int_1 h = \lim \int_1 h_n = \lim \int_2 h_n = \int_2 h.$$

This contradiction completes the proof that

2) $\int h$ exists for all $h \, \epsilon \, C^+$.

At the same time, because of the preceding argument, this

shows that if carrier $h_n \subset$ a fixed compact set independent of n, then

3) $\lim \int h_n = \int \lim h_n$

for all uniformly convergent sequences in C^+ (with carriers in W in the local group case). It is also true if h, h' and F_1 are in C^+ then

$$|h - h'| < \epsilon F_1 \text{ implies } |\int h - \int h'| < \epsilon \int F_1.$$

It follows also that if r is any real number and $f \epsilon C^+$ then

4) $r \int f = \int r f.$

We shall prove next for f, f' in C^+ that

5) $\int (f + f') = \int f + \int f'.$

The inequality $\int (f + f') \leq \int f + \int f'$ is known and it is only necessary to prove the reverse.

Let $F = f + f' + r F_1$, with $r > 0$, and where F_1 is a function in C^+ with positive lower bound on (carrier f) $\cup$ (carrier f') $= K$.

Let $h = f/F$ on K and zero elsewhere; similarly $h' = f'/F$ on K and zero elsewhere. Then F, h, $h' \epsilon C^+$, $hF = f$, $h'F = f'$, $h + h' \leq 1$ and $h + h'$ has carrier K.

Let $\epsilon > 0$ be arbitrary. There is a neighborhood V of e such that if $x \epsilon yV$ we have (by uniform continuity)

$$|h(x) - h(y)| < \epsilon, \text{ and } |h'(x) - h'(y)| < \epsilon.$$

Take i so large that

$$\text{carrier } g_i \subset W_i \subset V.$$

If

$$F(x) \leq \Sigma \, r_j g_i(a_j^{-1} x)$$

the term $g_i(a_j^{-1} x)$ will be zero except when $x \epsilon a_j V$. Since $h(x) \leq h(a_j) + \epsilon$ in those terms, we have also

$$F(x)h(x) \leq \Sigma' \, r_j g_i(a_j^{-1} x)(h(a_j) + \epsilon),$$

where Σ' is restricted to the terms not zero in $a_j V$ and therefore for all x

$$F(x)h(x) \leq \Sigma \, r_j g_i(a_j^{-1} x)(h(a_j) + \epsilon).$$

The same relation is true for h' and hence

$$(Fh : g_i) + (Fh' : g_i) \leqq \Sigma r_j(1 + 2\epsilon),$$

since $h + h' \leqq 1$. Since ϵ is arbitrary the right side may be made to approach Σr_j, and this may be made to approach $(F : g_i)$. Using this fact and remembering that $Fh = f$, $Fh' = f'$, we may conclude that

$$(f : g_i) + (f' : g_i) \leqq (F : g_i).$$

This shows that

$$\textstyle\int f + \int f' \leqq \int F$$

and so

$$\textstyle\int f + \int f' \leqq \int (f + f') + r\int F_1.$$

Since r is arbitrary 5) is proved.

More generally, combining 4) and 5) with r and s non-negative reals we have

6) $r\int f + s\int f' = \int (rf + sf')$.

The integral has so far been defined for C^+ but the definition can of course be extended to C. If $h \epsilon C$, then $h^+ = \mathrm{Max}\,(h, 0) \epsilon C^+$ and $h^- = -\,\mathrm{Min}\,(h, 0) \epsilon C^+$ and $h = h^+ - h^-$. Then $\int f$ can be extended to C in a natural way using the fact above so that 6) continues to hold for all real r and s. Then $\int f$ is a linear functional on C satisfying conditions 3), 4), 5), 6).

Take $f \epsilon C^+$ (for the case of a local group take $f \epsilon C^+$, carrier interior to W) and with the notation above suppose that

$$f(x) \leqq \Sigma r_j g_i(a_j^{-1} x)$$

for all x in G (in W^2). Let a be an element of $G(a \epsilon W)$. Then, letting $af(x) = f(a^{-1}x)$ we have

$$af = f(a^{-1}x) \leqq \Sigma r_j g_i(a_j^{-1}a^{-1}x)$$

for all $x \epsilon G(\epsilon W^2)$. Hence

$$\textstyle\int f(a^{-1}x) = \int f(x).$$

This property extends to $f \epsilon C$. Thus for f in C (carrier $f \subset W$), and $a \, \epsilon \, G \; (a \, \epsilon \, W)$ we have

    7) $\int af = \int f.$

This is the property of invariance of the integral.

2.18. Uniqueness of the integral

In this section and the next we consider only global groups. The results are valid for local groups but we omit them.

As mentioned above it was first shown by von Neumann [3] that the integral is essentially unique, that is, it is unique to within a scalar multiplier. We shall prove this theorem although we shall not have much need for it except to show that the integral exists in an arbitrary locally compact group, without using the Tychonoff Theorem.

Following A. Weil [1], it is convenient to use a construction due to Dieudonné which we now prove. Let K be a compact subset of a locally compact group G, let V be a compact neighborhood of e and suppose that every point of K is an inner point of $\cup y_i V$, $y_i \, \epsilon \, G$, $i = 1, \dots, n$. Let h_i' be a continuous function on G, $h_i' = 1$ on $y_i V$, $0 \leq h_i' \leq 1$, carrier $h_i' \subset y_i V^2$.

Let k be a continuous function, $k = 1$ on K, and $0 \leq k \leq 1$ everywhere. Now define the continuous function

$$m = \min \, (k, \; h' = \Sigma h_i'),$$

Then $m = 1$ on K and $m/h' \leq 1$. Finally define the partitioning functions

$$h_i = m h_i'/h'.$$

When $h' = 0$, m and $h_i' = 0$ and it is understood that $h_i = 0$. By construction the functions h_i are continuous, $h_i \geq 0$, and vanishing outside of $y_i V^2$ and $\Sigma h_i = 1$ on K.

THEOREM. *Let G be a locally compact group and C the class of continuous functions on G with compact carriers. If $\int^*$ denotes a non-trivial integral defined on C with properties 3), ..., 7) of the preceding section, then every such integral is of the form $r\int^*$ for some fixed real number $r > 0$.*

Note that this theorem does not use separability for G.

Let f and g belong to C^+, g not identically zero. Then for some real numbers r_j and elements a_j,

$$f(x) \leq \Sigma r_j g(a_j^{-1} x)$$

and therefore (letting $a_j g = g(a_j^{-1} x)$)

$$\int^* f \leq \Sigma r_j \int^* a_j g = (\Sigma r_j) \int^* g.$$

Since this is true for all Σr_j satisfying the preceding formula, it follows that

$$\int^* f \leq (f : g) \int^* g$$

where, as defined above, $(f : g) = $ g.l.b. (Σr_j).

Now let f be fixed, let K be the carrier of f and let $\epsilon > 0$ be preassigned. There is a $U(e)$ such that $y^{-1} x \epsilon U$ implies

$$| f(x) - f(y) | < \epsilon.$$

Choose any $g \epsilon C^+$ not identically zero with carrier in U and *symmetric*, that is, $g(x) = g(x^{-1})$. For x, $y \epsilon G$, x not in yU, $g(y^{-1} x) = 0$. For $x \epsilon yU$, $f(y) \geq f(x) - \epsilon$. Therefore for every x

$$\begin{aligned}
\int^* f(y) g(y^{-1} x) &\geq (f(x) - \epsilon) \int^* g(y^{-1} x) \\
&= (f(x) - \epsilon) \int^* g(x^{-1} y) \\
&= (f(x) - \epsilon) \int^* g.
\end{aligned}$$

Now let $\epsilon' > 0$ be arbitrary and choose $V'(e)$ so that $y \epsilon V' z$ implies

$$| g(y) - g(z) | < \epsilon'.$$

Let $V(e)$ be such that $V^2 \subset V'$ and choose elements y_i and partitioning functions $h_i \epsilon C^+$ such that $\Sigma h_i = 1$ on K, the carrier of f, and h_i vanishes outside of $y_i V'$. Then

$$\int^* f(y) g(y^{-1} x) = \Sigma_i \int^* f(y) h_i(y) g(y^{-1} x) dy.$$

But $h_i(y)$ is zero except for y in $y_i V'$ and $y \epsilon y_i V'$ implies $y_i^{-1} x \epsilon V' y^{-1} x$ and hence

$$g(y^{-1} x) \leq (g(y_i^{-1} x) + \epsilon').$$

Let $r_i = \int^* f h_i / \int^* g$; then $\Sigma r_i = \int^* f / \int^* g$. Combining the preceding inequalities

$$f(x) \leq \epsilon + \epsilon' \Sigma r_i + \Sigma r_i g(y_i^{-1} x).$$

Since ϵ' is arbitrary and not related to f and g,

$$f(x) \leqq 2\,\epsilon + \Sigma r_i g(y_i^{-1}x).$$

Assuming f not identically zero there are elements x_j, $j = 1, \ldots, N$, such that for $x \in K$,

$$1 \leqq \Sigma f(x_j^{-1}x).$$

Hence

$$f(x) \leqq \epsilon \Sigma f(x_j^{-1}x) + \Sigma r_i g(y_i^{-1}x).$$

Using the fact that $(x_j f : g) = (f(x_j^{-1}x) : g) = (f : g)$, it follows that

$$(f : g) \leqq \epsilon N(f : g) + \int^* f / \int^* g$$

Therefore

$$(f : g)(1 - \epsilon N) \leqq \int^* f / \int^* g.$$

It is important to observe that N does not depend on ϵ, the more so because $g(x)$ and therefore $\int^* g$ does not depend on ϵ.

Finally if f_0 is another non-trivial function in C^+ the same argument shows that

$$\frac{(\int^* f)(1 - \epsilon N_0)}{\int^* f_0} \leqq \frac{(f : g)}{(f_0 : g)} \leqq \frac{(\int^* f)}{(\int^* f_0)(1 - \epsilon N)}.$$

We see that if ϵ tends to zero and g is chosen in an appropriate carrier, the ratio $(f : g)/(f_0 : g)$ converges to $\int^* f / \int^* f_0$. It follows that the ratio $\int^* f / \int^* f_0$ depends on the functions f and f_0 and not on $\int^*$. This completes the proof.

2.19. Integrals on non-separable groups

Let G be a locally compact group and H a compact invariant subgroup. Suppose that an integral is defined on G. Let f' be a continuous function on G/H, with compact carrier. Then there is a function f on G uniquely defined by

$$f(x) = f'(xH),$$

which is continuous and has a compact carrier. Define an integral on G/H by

$$\int' f' = \int f,$$

where the first integral is on G/H and where the second integral is on G. The functions f range over a linear family of functions on G closed under uniform convergence and the integral restricted to them has its usual properties that is 3)—7) of 2.17. It follows that $\int' f'$ is the integral on G/H to within a scalar multiplier.

Now let G be a locally compact group, G' an open (hence closed) subgroup as described in 2.3.1 with G'/G_0 compact. Let $f_0 \in C^+$, $f_0 \not\equiv 0$, where C is the class of continuous functions with compact carriers in G'. We shall define a left invariant integral on G' so that $\int^* f_0 = 1$. Take $f \in C$, $f \not\equiv 0$. There is an appropriate subgroup H, that is a compact invariant H such that f and f_0 are constant on cosets of H, and G'/H is separable metric.

Let f_0' and f' be the functions on G'/H determined by f_0 and f. We define the integral of f on G' by

$$\int^* f = \int' f'$$

for that unique integral $\int'$ on G'/H for which

$$\int' f_0' = 1.$$

We have to verify that $\int^* f$ does not depend on the particular choice of H.

It is sufficient to verify that if H' is appropriate to f and f_0 and $H' \subset H$, then $\int^* f$ is the same number whether defined from H or from H'. But in this case we have the natural maps

$$T' : G' \to G'/H'; \quad T_1 : G'/H' \to G'/H$$

and

$$T = T_1 T' : G' \to G'/H.$$

It follows from the opening paragraphs of this section that with the normalization

$$\int' f_0' = 1; \quad \int'' f_0'' = 1$$

the first integral on G'/H, the second on G'/H', the following

equality holds
$$\int' f' = \int'' f''.$$
This shows that $\int^* f$ is well defined. It is a fact that given any countable set of functions of C there is at least one H which is appropriate to all of them (2.8); it follows that $\int^* f$ enjoys all the properties of an integral including convergence for uniformly converging sequences with carriers in a fixed compact set.

Finally, it is a simple matter to extend this integral over all of G. First if V is a compact subset of G then V is the union of a *finite* number of disjoined compact sets each of them contained in a unique coset of G/G'. Then the integral of a function with carrier in V will be defined as the sum of the integrals over the individual cosets. Next if f is a function with carrier V in a coset aG', for some $a \,\epsilon\, G$, then we define
$$\int_G f(x)dx = \int_{G'} f(a^{-1}x)dx.$$
It follows from the left invariance of the integral in G' that this definition is unique.

2.19.1. We conclude this discussion of the integral with some remarks on the iterated integral. Let $f(x, y)$ be a function on G which is continuous in x and y and vanishes when either x or y is outside of some compact set U. Then $\int f(x, y)dy$ is defined for each value of x and is continuous in x, and vanishes outside of U. Therefore the iterated integral $\int\int f(x, y)dydx$ exists and it defines an integral on the product group $G \times G$. Since $\int\int f(ax, ay)dydx = \int\int f(ax, v)dvdx = \int\int f(u, v)dvdu$, it follows that this integral is left-invariant on $G \times G$.

Similarly, the iterated integral $\int\int f(x, y)dxdy$ defines a left-invariant integral on $G \times G$. Both integrals have the standard properties, and therefore the two integrals coincide provided they give the same value for at least one non-trivial function. It is obvious that the integrals agree for symmetric functions, and therefore the two iterated integrals are equal.

We shall soon refer to some of the classical results on characteristic functions of symmetric kernels; the use of properties of the iterated integral is implicit in that theory.

2.20. Compact groups

The existence of an invariant integral in Lie groups can be established in a simple and direct way. In 1929, Peter and Weyl [1] introduced the methods of integral equations in the study of compact Lie groups. Their methods were carried over to the general compact groups by von Neumann once an invariant integral was available. Only the beginning of their theory is needed to show that compact groups have representations as subgroups of matrix groups Gl_n. In the general case, no single representation can be an isomorphism; but to every neighborhood U of the identity there is a representation whose kernel lies in U. In the next paragraph we shall formulate *without proof* the part of the theory of characteristic functions which is needed for the present application. The proofs are available in many standard works of analysis, Courant-Hilbert for example. The complete details will also be found in Chevalley ([1], pp. 204—212); nothing in those pages depends on special properties of Lie groups.

Let G be a compact group with a left-invariant integral defined on it (it is not important for the sequel that for compact groups the same integral is also right invariant). Let $f(x)$ be a real continuous function on G and let $k(x, y)$ be a real function continuous on $G \times G$. Then

1) $g(x) = \int k(x, y) f(y) dy$

is a continuous function of x. We shall assume that the kernel $k(x, y)$ is symmetric, i.e. $k(x, y) = k(y, x)$.

Any non-trivial continuous function on G satisfying

2) $f(x) = c \int k(x, y) f(y) dy$

is called a *characteristic function* for the number c; and c, which is necessarily real, is called a *characteristic number*. *The characteristic functions for the same c form a finite-dimensional linear space.*

Characteristic numbers and their associated functions exist for every choice of a symmetric continuous kernel on G. Moreover, if a function $g(x)$ *can be represented as a transform* 1) *of a con-*

tinuous function $f(x)$, *then* $g(x)$ *can also be represented as the sum of a uniformly convergent sequence of characteristic functions of the kernel* k:

3) $g(x) = f_1(x) + f_2(x) + \ldots;$

the functions $f_n(x)$, satisfying relation 2).

We shall show now that if a kernel k has the special form $h(x^{-1}y)$ for a continuous function $h(z) = h(z^{-1})$ and if c is a characteristic number for the kernel $h(x^{-1}y)$, then the characteristic functions for the number c are permuted by the elements of G. Let $f(x)$ satisfy

4) $f(x) = c \int h(x^{-1}y)f(y)dy$

and let a be an element of G. Then

$$af(x) = f(a^{-1}x) = c \int h(x^{-1}ay)f(y)dy = c \int h(x^{-1}z)f(a^{-1}z)dz,$$

and we see that $af(x)$ is a characteristic function for the number c, as asserted.

It is clear that G is a linear transformation group of the space of continuous functions on G and therefore for each characteristic number c we obtain a representation of G as a linear group of transformations of a finite-dimensional space. We prove next that if b is an arbitrary element of G, $b \neq e$, we can find a representation of G in which b is not in the kernel: i.e. b is not mapped into the unit matrix of Gl_n. Let U be a symmetric neighborhood of e such that

$$b \notin U^2.$$

Let f be a continuous non-negative function vanishing outside of U and equal to 1 at e. Let $h(z) = f(z) + f(z^{-1})$. Now define a function g by

5) $g(x) = \int h(x^{-1}y)f(y)dy.$

Then $g(e) \geq \int (f(y))^2 dy > 0$ and, at the same time, g vanishes outside of U^2. Therefore it follows that

6) $bg \neq g.$

Now let us write g in the form 3) for appropriate characteristic functions f_1, f_2, ..., for the present kernel h. It follows since 3) holds for every x that

7) $bg = bf_1 + bf_2 + \ldots$

It follows from 6) that for at least one characteristic function f_m,

$$bf_m \neq f_m.$$

This function gives the desired representation of G as a subgroup of Gl_n, where n is the number of linearly independent characteristic functions which correspond to the same characteristic number as the function f_m. The kernel of this map of G into Gl_n is the set of elements for which $af_m = f_m$, and b is not among them.

With this background it is easy to prove the principal result of this section.

THEOREM. *Let G be a compact group and let U be an open neighborhood of e. Then U contains an invariant subgroup H of G such that G/H is isomorphic to a subgroup of some Gl_n, and in fact to a subgroup of the orthogonal group of some euclidean space.*

Let a denote an arbitrary element of G not in U. Let H_a denote an invariant subgroup of G not containing a such that G/H_a is a representation. Since H_a is closed, and the complement of U is compact, there is a finite set of such subgroups:

$$H_1, \ldots, H_n,$$

whose intersection is contained in U.

Let E_i, $i = 1, \ldots, n$, denote euclidean spaces of appropriate dimensions such that G/H_i is isomorphic to a linear transformation group of E_i. If we form the product space

$$E = E_1 \times E_2 \times \ldots \times E_n,$$

we can define the action of each element of G upon E from its known action on each factor E_i. In this way G becomes a linear transformation group of E whose kernel is the intersection of the given H_i; the kernel is contained in U as asserted.

The fact that G can be regarded as an orthogonal group on each space E_i, and therefore on the product space E, arises from the fact that G is an *orthogonal* group on the space of continuous functions on G. This means that G leaves invariant a *scalar-product* of functions:

$$(f, g) = \int f(x)g(x)dx,$$

leading to a norm $\| f \| = (f, f) \geqq 0$. Thus

$$\begin{aligned}(af, ag) &= \int f(a^{-1}x)g(a^{-1}x)dx \\ &= \int f(z)g(z)daz \\ &= \int f(z)g(z)dz = (f, g).\end{aligned}$$

For a euclidean space, the existence of this invariant scalar product is equivalent to the invariance of a positive definite quadratic form, or equivalently a distance-function.

COROLLARY. *There is an open neighborhood W, $U \supset W \supset H$ such that any subgroup of G which is in W is a subgroup of H.*

This follows from the fact that G/H has no small subgroups, since it is contained in Gl_n which has none. Thus let W^* be a neighborhood of the unit in G/H and let W^* have no subgroups. Let T denote the map of G onto G/H, and let $W = U \cap T^{-1}W^*$.

The preceding proposition will be of the utmost importance in the sequel. The following proposition is of interest.

COROLLARY. *A compact group G is isomorphic to a subgroup of the product of a number (possibly uncountably infinite) of orthogonal groups of appropriate dimensions.*

Take a basis for neighborhoods at e. With each neighborhood U_c of e, associate a subgroup H_c in U_c such that G/H_c is a subgroup of some orthogonal group O_c. Let O^* denote the product-space (1.6) and product group of all the O_c. Each element x of G has a coordinate x_c in the "factor" O_c, namely the element of O_c upon which it is mapped. If x is not e, then for at least one value of c, x is not in H_c, and x_c is different from e_c. This shows that the mapping of G into O^* is one-one. Since the mapping is continuous one way, and the spaces are compact, the map is a homeomorphism. Therefore the map is an isomorphism, and the theorem is proved.

The analysis of compact groups, only a small part of which has been shown here, was achieved by von Neumann in 1933.

2.20.1.   We have shown that if a compact group does not have arbitrarily small subgroups, that is if there is a neighborhood U of e such that e is the only subgroup of G contained in U, then G is isomorphic to a subgroup of some Gl_n and, consequently, G is a Lie group.   We shall see later that if a locally compact group does not have small subgroups it is necessarily a Lie group. However it is known that a Lie group G is not necessarily isomorphic to a subgroup of some Gl_n, if G is not compact. It is true but will not be considered in this book that every Lie group is locally isomorphic to a subgroup of some Gl_n.

2.21. Abelian groups

The analysis of locally compact abelian groups can be made to depend upon the fact, proved by Pontrjagin in 1934, that they are locally isomorphic to compact groups.

THEOREM. *Let G be a locally compact abelian group. Let U be a compact symmetric neighborhood of e. Let G' be the subgroup of G generated by U.   Then G' contains a discrete subgroup D with a finite number of generators such that G'/D is compact.*

In any group G' which is generated by a compact neighborhood U, whether or not G' is abelian, it is easy to obtain a subgroup A with a finite number of generators such that

$$G' \subset AU.$$

For, since U^2 is compact there is a finite set of elements, $a_1\ a_2, \ldots, a_n$ of G' such that $U^2 \subset \cup a_i U$. Let A be the group generated by the elements a_i.   Then

$$U^2 \subset AU$$
$$U^3 \subset AU^2 \subset AAU = AU,$$

and so by induction $U^n \subset AU$, for every n. Therefore $G' = \cup U^n \subset AU$ as asserted.

Except in the abelian case, the group A will not need to be invariant; and in even the present case that G' is abelian the group A is not necessarily discrete. To find the desired *discrete* subgroup D of A, such that $G' \subset DU$ we need the following Lemma.

LEMMA. *Let T denote the additive group of integers (the same result and the same proof hold if T is the additive group of reals). Let f be a homomorphic map of T into a locally compact group G (not necessarily abelian). Then, either this map is an isomorphism or $f(T)$ is contained in a compact subgroup of G.*

Let K denote the closure of the group $f(T)$; K is a closed abelian subgroup of G (1.11.1) and we now confine our attention to K. Suppose that there can be found a neighborhood U' of e (in K) and also a positive integer n such that $U' \cap f(T)$ contains *no* element $f(t)$ for which $t \geq n$. In that case it is easy to see that $f(T)$ is an isomorphism. We shall suppose that $f(T)$ is not an isomorphism, and then every (symmetric) neighborhood of e meets $f(T)$ in elements $f(t)$ for arbitrarily large $t > 0$. We must now show that K is compact.

Let V be an open symmetric neighborhood of e in K such that $\overline{V}$, the closure of V, is compact. Since $f(T)$ is dense in K, every element x of K is contained in some $f(t)V$ and we may suppose that $t > 0$, in view of the remarks above. Because $\overline{V}$ is compact, there is a finite set of positive numbers $t_1, \ldots, t_n$ of T such that $\overline{V} \subset \cup_i f(t_i)V$. Let t^* be the largest number among the t_i.

Now let x be an arbitrary element of K and define $t(x) = t \varepsilon T$ as the smallest non-negative number such that $x \varepsilon f(t)\overline{V}$. Because of the symmetry of V, we may write $f(t)x^{-1} \varepsilon \overline{V}$ and there is at least one of the elements t_i, above, such that $f(t)x^{-1} \varepsilon f(t_i)\overline{V}$. Then

$$x^{-1} \varepsilon f(t_i - t)\overline{V},$$
$$x \varepsilon f(t - t_i)\overline{V}.$$

It follows from the definition of $t(x)$ that $t - t_i \leq 0$. Therefore $t(x) \leq t^* = \max. (t_i)$. Since x is arbitrary in K it follows that $K \subset f(T^*)\overline{V}$ where T^* is the interval on T from $t = 0$ to $t = t^*$. Thus K is compact, and the Lemma is proved.

We return to the main theorem and the interrupted proof. Suppose that each a_i of A generates a compact subgroup. In that case it follows that A is contained in the set of products of elements of those groups, and A is contained in a compact set (2.1.1). But then $G' = AU$ is compact and there is nothing more to prove.

Accordingly we now suppose that the elements $a_1, \ldots, a_n$ generating A have been so arranged that the group generated by the first $k \leq n$ is a discrete group D and that if $k < n$ then for each i, $k < i \leq n$ the element a_i and the group D generate a closed subgroup F_i which is not discrete. If $k = n$, then $D = A$ and G'/A is compact; the Lemma is proved.

We now suppose that $k < n$. For each i, $k < i \leq n$, F_i is not discrete and F_i/D which is locally isomorphic to it is not discrete. The map $F_i \to F_i/D$ takes $\{a_i^m\}$ to a dense set in F_i/D and it follows from the Lemma above that F_i/D is compact. By the argument which was used in 2.3.1·it follows that F_i contains a compact set W_i such that $F_i = DW_i$.

Now let W denote the set-product of the compact sets W_i for all $i = k + 1, \ldots, n$, and consider the set DWU. Every element of A is contained in the product $DF_{k+1} \ldots F_n$. But

$$DF_{k+1} \ldots F_n \subset D(DW_{k+1}) \ldots (DW_n)$$
$$= DW_{k+1} \ldots W_n$$
$$= DW.$$

Therefore $A \subset DW$. Then $G' \subset AU \subset DWU$. But WU is compact (2.1.1) and therefore G'/D is compact. This proves the Theorem.

COROLLARY. *The group D may be assumed to contain no elements of finite order.*

We prove this by arranging the generators of D so that the first h, $h \leq k$, generate a group D' with no element of finite order, but so that any remaining generator·(if $h < k$) introduces elements of finite order. Then for each i, $h < i \leq k$, we get a group F_i such that F_i/D' is *finite*. The rest of the proof is as above.

There is an arbitrarily small subgroup H' of G'/D such that $(G'/D)/H'$ is a compact subgroup of some Gl_n. Since D is discrete we may choose H' small enough so that it is isomorphic to a subgroup H of G'; i.e. there is a subgroup H of G' such that G'/DH is a compact Lie group (DH is the group in G' which is the product of D and H). From this point of view it is clear that G'/H is a group locally isomorphic to the compact Lie group G'/HD. We have proved the following corollary to the theorem.

COROLLARY. *There is an arbitrarily small compact subgroup H of G' such that G'/H is locally isomorphic to a matrix group and therefore is a Lie group.*

2.21.1. One of the most interesting theorems about abelian groups is the following by Pontrjagin.

THEOREM 1. *Let G be a locally compact connected abelian group. Then G contains a maximal compact subgroup K and G/K is an r-dimensional linear vector-space.*

Since G is connected it may be generated by a neighborhood of the identity and therefore the preceding corollary applies to it. Let H be a compact subgroup such that G/H is locally isomorphic to V_n. By the Theorem at the conclusion of 2.16 we may write $G/H = V_r \times T_k$, where T_k is a k-dimensional toral group. The inverse-image of T_k in G is a compact subgroup K such that G/K is isomorphic to V_r. This completes the proof.

COROLLARY. *Let G be an abelian locally compact group. Then G contains an open subgroup G'' and G'' contains a compact subgroup H^* such that G''/H^* is a V_r.*

Let U be a compact neighborhood of e and let it generate the open subgroup G' of G. Then G' has a compact subgroup H such that G/H is *locally* isomorphic to a V_n. Now the identity-component of G/H is the product of a V_r and a toral T_k. Let G'' be the inverse image of the identity-component of G/H and let H^* be the inverse image of T_k. This completes the proof.

It is not difficult to show that G'' contains a subgroup isomorphic to V_r such that $G'' = H^* \times V_r$, and also to show that V_r is a direct factor of the group G. We shall postpone this to Chapter 4 because the operation of "lifting" an n-cell from a factor-group back into the group will come up again in a more general setting.

2.22. Existence of one-parameter subgroups

We conclude this chapter with the proof of two special cases of the following proposition: *a locally compact metric connected (non-trivial) group G has a (non-trivial) one-parameter local subgroup $x(t)$.* Of course, each local group is contained in a unique extension $x(t)$ defined for all t but the extended group is not necessarily closed.

Case 1). *Let G be compact.* Let $H_1 \supset H_2 \supset \ldots$, be a sequence of invariant subgroups of G *converging* to e such that G/H_n is a matrix group, $n = 1, 2, \ldots$. Let $x_1(t)$ be a (non-trivial) local one-parameter subgroup of G/H_1. Let T denote the natural map of G/H_2 upon G/H_1. Let U_2 be a neighborhood of e in G/H_2 such that every element is on a unique one-parameter local group contained in U_2. Then $U_1 = T(U_2)$ is a neighborhood of e in G/H_1 and U_1 contains an element $x_1(t')$ *distinct from e* (for sufficiently small t'). Let $x_2(t')$ be an element of U_2 covering $x_1(t')$ and let $x_2(t)$ be the one-parameter local group in G/H_2 which contains $x_2(t')$. It is clear that $T(x_2(t)) = x_1(t)$ for all values of t for which both groups may have been defined.

Continuing in this way we can construct groups $x_1(t)$, $x_2(t)$, $\ldots$, in $G/H_1, G/H_2, \ldots$, respectively, each projecting upon the preceding, and all defined for the same range of t, say $|t| \leqq 1$. To these groups there corresponds the monotonic sequence of compact sets: $x_1(t)H_1$, $x_2(t)H_2$, $\ldots$, in G, and the intersection of these sets is the desired local group $x(t)$.

Case 2). *Let G be abelian and locally compact.* Then G contains a discrete group D, and G/D is compact and connected and contains a one-parameter local group $x(t)$. Since G and G/D are locally isomorphic, this gives a one-parameter group in G.

In the next chapter we shall conclude the demonstration of the *general* proposition (p. 118).

CHAPTER III

Groups With No Small Subgroups

3.0. Introduction

Every topological group G may be regarded as a transformation group over itself as space through its left (and right) translations: $x \to gx$ (and $x \to xg$) and through the taking of transforms:

    1) $x \to g^{-1}xg$.

The translations exhibit G as an *effective* transformation group, but in the case of 1) the elements of the center of G are inoperative and the effective transformation group is G/H, where H is the center of G.

A Lie group being a differentiable manifold has associated with it at each point a tangent-space; these are vector-spaces of the same dimension, say n, as the group. The group operations of a Lie group are differentiable, and therefore they *permute* the tangent-spaces. In particular, the transformations 1) carry the tangent-space at the identity into itself and in this way our group becomes an automorphism group of an n-dimensional vector-space. This is called the *adjoint-representation* of a Lie group.

Every Lie group has neighborhoods of the identity that are uniquely filled by one-parameter local groups radiating to the boundary; each one-parameter group corresponds to an initial direction-vector. In $M_n(R)$, the additive group of real $n \times n$ matrices, the whole group is such a neighborhood. More generally, neighborhoods of this sort can be obtained in any closed subgroup of $Gl\ (n,\ R)$ by the use of the map of M_n into Gl_n:

    2) $A \to \exp A$,

which, as we saw in 2.16, is a local homeomorphism carrying one-parameter groups into one-parameter groups. To each closed subgroup of Gl_n, 2) makes correspond an appropriate linear subspace of M_n (2.16).

The formula which was very important before,

3) $\exp (A + B) = \lim_{n \to \infty} (\exp A/n \exp B/n)^n,$

shows that to the operation of addition in the vector-space associated with a subgroup G of Gl_n there corresponds an intrinsically defined operation in G. If $x(t)$, $y(t)$, are local one-parameter subgroups of G in a sufficiently small neighborhood of e, then 3) enables one to define an operation of "addition" in G:

4) $x(t) + y(t) = \lim_{n \to \infty} (x(t/n)y(t/n))^n.$

In terms of this new operation the set of elements on one-parameter local groups in a neighborhood of e in G becomes a local linear vector space. We shall prove this *in a much more general setting* in 3.9, 3.10 of this chapter.

It is not difficult to see that the transformations 1) "commute" with the operation of addition defined by 4) and in this way one is led again to the adjoint representation of the group G. This will be generalized in 3.11 of this chapter.

A Lie group has no small subgroups; that is, there is a neighborhood of e which has no other subgroup than e. Conversely, *if G is a locally compact group without small subgroups then G is a Lie group.* This result was proved by Gleason [5] (1952) for finite-dimensional groups and the result was extended by Yamabe [3] (1953). Not quite all of this will be proved in the present chapter. We show here that a group G with no small subgroups has an "adjoint representation": that is, by means of transformations 1) G becomes homomorphic to a subgroup of Gl_n. At that point we shall know that the kernel H of the representation and the factor group G/H are both Lie groups.

If $x(t)$, $y(t)$ are one-parameter local groups in a group then $z_n = (x(1/n)y(1/n))^n$ can be defined for every n. But it is not obvious that z_n converges as n goes to infinity, and it is not even obvious that z_n remains in a fixed compact subset (in which case some *subsequence* will converge). The first sections, following Gleason and the improvements by Yamabe, set up the machinery for proving the boundedness of z_n. They are used again in the next chapter for the general case (small subgroups being present).

In 3.4 to 3.6 we construct the basic family of one-parameter local groups of G to which 4) is then to be applied. In 3.7 we prove the uniform boundedness of z_n when $x(1)$, $y(1)$ are in a suitable compact neighborhood of e. In 3.8 we prove a basic Lemma from which one derives the uniform continuity of the new addition-operation: $x + y$. That we obtain a *finite-dimensional* representation follows. The methods in Chapters III and IV derive principally from Gleason [5] and Yamabe [3, 4].

An important partial result along these same lines not restricted to finite-dimensional groups had been found earlier by Kuranishi (1949) and we follow him in some details. This is particularly true of our replacing the explicit use of a "tangent-space" by the device of defining a new "addition-operation" in the given group. This seems to lead naturally to certain small differences in connection with the use of the elements of G as "operators" (3.3).

In the next chapter, in order to analyze the general locally compact group we shall need to use the fact that compact groups can be approximated by Lie groups (2.20). It is interesting that one can study the class of groups without small subgroups without first "sorting out" the compact case.

3.1. Generating sequences of sets

Throughout this chapter G will denote a locally compact group which satisfies the first countability axiom; as we saw in 1.22 this implies that G is metric. We shall not make any systematic use of a particular metric on G, but a metric is used implicitly in connection with properties of sequential convergence of sets (1.10). We remind the reader that the notation AB designates the product of sets; A^n is the collection of products of n elements, each of them in A, A^0 is the identity. The symbol $[rk]$ as an exponent denotes the largest integer $\leq rk$.

Suppose that A_n, $n = 1, 2, \ldots$, is a sequence of somehow defined compact symmetric subsets of G converging to e, $e \varepsilon A_n$, $A_n \neq e$:

1) $\lim A_n = e$.

Let U denote some compact neighborhood of e and assume that *for each n there exists some power of A_n which is not contained in U.* In that case there will exist a sequence of integers $k(n)$ such that

2) $A_n^{k(n)+1} \not\subset U$ but $A_n^i \subset U$ for $i \leq k(n)$;

the integers $k(n)$ necessarily increase without limit because of 1).

It follows from 1.10 that there is a subsequence $I' \subset I$ such that $\lim A_n^{k(n)}$ exists and also $\lim A_n^{[k(n)/2]}$. Let

3) $\lim A_n^{k(n)} = E(1)$, $\lim A_n^{[k(n)/2]} = E(\tfrac{1}{2})$, $n \in I'$;

then,

4) $e \in E(\tfrac{1}{2}) \subset E(1) \subset U$.

LEMMA. *The set $E(1)$ contains a point of the boundary of U.*

Let q_n denote a point of $A_n^{k(n)+1}$ which is *not* a point of U; $q_n = p_n x_n$ where x_n is some point of A_n and p_n is in $A_n^{k(n)}$. The sequence q_n belongs to U^2 and U^2 is compact: therefore there is a convergent subsequence. Let q denote a limit point of such a convergent subsequence. Since x_n converges to e, q is also a limit point of the sequence p_n and belongs to $E(1)$. This completes the proof.

COROLLARY. *The set $E(1)$ is connected.*

Suppose that $E(1)$ is the union of compact non-empty non-intersecting sets A and B, and that $e \varepsilon A$. There is a compact neighborhood V of e such that $AV \cap BV = \Phi$ (2.1). It is clear that no point of $E(1)$ is on the boundary of AV. Since B is not empty, it is *not* possible that all the sets A_n, A_n^2, $\ldots$, $A_n^{k(n)}$ belong to AV for *arbitrarily large n.* This permits us to find an appropriate subsequence of the sets A_n and appropriate powers $\leq k(n)$ which generate a set $E'(1) \subset AV$ meeting the boundary of AV. Since $E'(1)$ is in $E(1)$, this is a contradiction.

3.1.1. LEMMA. *Let A_n, $n \in I' \subset I$, be compact symmetric sets such that* $\lim A_n = e$, *and let them generate the sets* $E(1)$ *and* $E(\frac{1}{2})$: $\lim A_n^{k(n)} = E(1)$, $\lim A_n^{[k(n)/2]} = E(\frac{1}{2})$. *Suppose that* $E(1)$ *is not a group and let* p *be a point of* $E(1)^2 - E(1)$. *Then*

$$pE(\tfrac{1}{2}) \cap E(\tfrac{1}{2}) = \Phi.$$

Of course, if $E(1)^2 - E(1)$ is empty then $E(1)$ is a group. Therefore it is not empty and contains some point p. Suppose that $px = y$, where x and y belong to $E(\frac{1}{2})$. Then $x = \lim x_n$ and $y = \lim y_n$ for some sequence of points x_n, y_n in $A_n^{[k(n)/2]}$. Now $p = yx^{-1} = (\lim y_n)(\lim x_n^{-1}) = \lim y_n x_n^{-1} \in \lim A_n^{k(n)} = E(1)$. The contradiction completes the proof.

COROLLARY. *Under the hypotheses above, there exists a compact neighborhood X of e such that*

$$pE(\tfrac{1}{2})X \cap E(\tfrac{1}{2})X = \Phi,$$

and such that for all sufficiently large $n \varepsilon I'$ (as above)

$$pA_n^{[k(n)/2]} X \cap A_n^{[k(n)/2]} X = \Phi.$$

The first relation follows from the Lemma and from 2.1. The second relation follows from the first and the definition of sequential convergence (1.10).

3.1.2. It is extremely interesting, although we shall not need to take advantage of it, that sets $E(r)$ can be defined by the diagonal process for every rational r and that they form a semigroup of sets:

$$E(r)E(s) = E(r + s).$$

The set $E(0)$ is defined separately as $\cap E(r)$ for all $r > 0$; it is a compact subgroup contained in U. When $E(0) = e$, the definition of these sets can be extended easily to all positive reals, and one obtains a *one-parameter semigroup of compact sets* (Gleason [2, 5]). We shall not pursue the details here. An interesting study of the kinds of semigroups which may be constructed on Lie groups is in Rädstrom (Arkiv för Mat, vol. 2).

3.2. Smoothing functions: Δ_n

We continue to use the sets A_n, of 3.1 defined for all $n \varepsilon I$. For each n, the function $\Delta_n(x)$, $x \varepsilon G$, is defined as follows:

$$\Delta_n(x) = 1 \text{ if } x \text{ is not in } A_n^{[k(n)/2]},$$
$$= 2j/k(n) \text{ if } x \varepsilon A_n^j - A_n^{j-1}, \ j \leq k(n)/2, \ (A_n^0 = e)$$
$$= 0 \text{ if } x = e.$$

We see that $\Delta_n(x) = 0$ if and only if $x = e$. These functions are *symmetric* and *lower semicontinuous* and satisfy the following:

1) $\quad \Delta_n(xy) \leq \Delta_n(x) + \Delta_n(y)$

for all x, y.

Let $\theta_0(x)$, $\theta_0(x) \varepsilon [0, 1]$ be a fixed continuous function with an arbitrary compact carrier $X \subset U$, such that $\theta_0(x) = 1$ if and only if $x = e$. For each n define the *smoothed* function θ_n,

2) $\quad \theta_n(x) = \text{l.u.b.}_y \ (1 - \Delta_n(y))\theta_0(y^{-1}x).$

Since $\theta_0(y^{-1}x)$ is zero when $y^{-1}x$ is not in X, and $1 - \Delta_n(y) = 0$ when y is not in $A_n^{[k(n)/2]}$, it follows that $\theta_n(x)$ vanishes for x outside of $A_n^{[k(n)/2]}X$. For all x and all n

$$0 \leq \theta_0(x) \leq \theta_n(x) \leq 1.$$

We shall show next that the functions $\theta_n(x)$ are continuous and in fact equicontinuous. Let $\epsilon > 0$ be given. Since θ_0 is continuous on a compact carrier there exists a symmetric $V_\epsilon(e)$ such that for all y

3) $\quad | \theta_0(yx) - \theta_0(yz) | < \epsilon$

provided $x^{-1}z \varepsilon V_\epsilon$. Let an x and z be related in this way. By definition

$$\theta_n(z) = \text{l.u.b.}_y(1 - \Delta_n(y)) \theta_0(y^{-1}z).$$

For this fixed z,

$$(1 - \Delta_n(y))\theta_0(y^{-1}z)$$

regarded as a function of y is upper semicontinuous on a compact carrier in U and therefore there is a $\bar{y}$ depending on z such that

4) $\theta_n(z) = (1 - \Delta_n(\bar{y}))\theta_0(\bar{y}^{-1}z)$.

For the same element $\bar{y}$,

$$\theta_n(z) - \theta_n(x) \leq (1 - \Delta_n(\bar{y}))\,(\theta_0(\bar{y}^{-1}z) - \theta_0(\bar{y}^{-1}x)) \leq \epsilon.$$

This relation holds for any x and z for which $x^{-1}z \,\epsilon\, V_\epsilon$. But $x^{-1}z \,\epsilon\, V_\epsilon$ implies $z^{-1}x \,\epsilon\, V_\epsilon$ and it follows that, *independently* of n, including $n = 0$,

5) $|\,\theta_n(z) - \theta_n(x)\,| \leq \epsilon.$

This completes the proof of the equicontinuity of the family of functions θ_n.

In the next section we derive from 5) certain important formulas.

3.2.1. We are going to interpret the elements of U as *operators* on the functions θ_n: if $a \,\epsilon\, U$, then $a\,\theta_n(x) = \theta_n(a^{-1}x)$; we will not always indicate the independent variable, writing merely $a\theta_n$.

What is more likely to cause confusion is that when a, b are used as operators in succession on the same function then

$$abf(x) = af(b^{-1}x) = f(b^{-1}a^{-1}x)$$

as it should since

$$(ab)^{-1} = b^{-1}a^{-1}.$$

To see this more clearly the reader will remember that $bf(x)$ is a new function, say g, which has that value at the place bx that f has at the place x (for every x). Then, $abf(x) = ag(x)$ has that value at the place abx which g has at the place bx.

We can compare two elements a, b of U by forming the new function

$$(a - b)\theta_n = a\theta_n - b\theta_n.$$

This leads us to consider expressions of the form

$$\theta_n(ay) - \theta_n(by)$$

for a range of values of y, and these do not quite answer to 5), as we shall explain below.

Let $V_\epsilon(e)$ be given. A neighborhood V'_ϵ of e exists (2.4) such that for every $y \, \epsilon \, U^2$,

$$y^{-1} V'_\epsilon y \subset V_\epsilon.$$

Now it follows from 3.2 that given any $\epsilon > 0$, there exists a $V'_\epsilon(e)$ such that if $a^{-1}b \, \epsilon \, V'_\epsilon$ then for all y in U^2

6) $| \, \theta_n(ay) - \theta_n(by) \, | < \epsilon.$

It will be convenient to have a neighborhood V^*_ϵ such that 6) continues to hold if $a^{-1}b \, \epsilon \, V^*_\epsilon$ and *also* if $ab^{-1} \, \epsilon \, V^*_\epsilon$. Since $(a^{-1}b)^{-1} = b^{-1}a \neq ab^{-1}$, in general, one has to keep in mind that $b^{-1}a$ may be "near" e and ab^{-1} may not be near e. However, if a, b are restricted to a compact set, say U_0^2, then given $V(e)$ there exists $V^*(e)$ such that:

7) $ab^{-1} \, \epsilon \, V^*$ implies $a^{-1}b \, \epsilon \, V.$

Recall that U_0 has first countability. If 7) were not true there would exist sequences a_n, b_n such that $a_n b_n^{-1}$ converged to e, and such that $a_n^{-1} b_n \, \epsilon \, V$. Now, if a_n converges to a and b_n to b, then $a = b$. But then $a_n^{-1} b_n$ converges to e, and this is what is needed for a proof by contradiction.

3.2.2.   Therefore we may sum up in the formula: given any $\epsilon > 0$ there is a symmetric $V^*_\epsilon(e)$ such that if a, b, $y \, \epsilon \, U_0^2$ and if $a^{-1}b \, \epsilon \, V^*_\epsilon$ *or if* $ab^{-1} \, \epsilon \, V^*_\epsilon$ then independently of y and for all n including $n = 0$

8) $| \, \theta_n(ay) - \theta_n(by) \, | < \epsilon.$

3.2.3.   The following functional relation is a generalization by Yamabe [4] of a relation which Gleason [5] introduced as a type of Lipschitz condition.

LEMMA.   *For all* $n \, \epsilon \, I$ *and all* $a \, \epsilon \, G$, *independently of* $x \, \epsilon \, G$

$$| \, \theta_n(ax) - \theta_n(x) \, | \leq \varDelta_n(a)$$

By definition,

$$\theta_n(ax) = \text{l.u.b.}_y \, (1 - \varDelta_n(y)) \theta_0(y^{-1}ax)$$
$$= \text{l.u.b.}_z \, (1 - \varDelta_n(az)) \theta_0(z^{-1}x).$$

As remarked in connection with 4) there exists an element $\bar{z}$ depending on x, such that

$$\theta_n(x) = (1 - \Delta_n(\bar{z}))\theta_0(\bar{z}^{-1}x).$$

For the same x and $\bar{z}$

$$\theta_n(ax) \geq (1 - \Delta_n(a\bar{z}))\theta_0(\bar{z}^{-1}x)$$

and

$$\theta_n(x) - \theta_n(ax) \leq (\Delta_n(a\bar{z}) - \Delta_n(\bar{z}))\theta_0(\bar{z}^{-1}x)$$
$$\leq \Delta_n(a)\theta_0(\bar{z}^{-1}x)$$
$$\leq \Delta_n(a).$$

Since the relation holds for all a and since the right hand side is independent of x we may use $a^{-1}ax = x$ to write

$$\theta_n(ax) - \theta_n(x) \leq \Delta_n(a^{-1}) = \Delta_n(a).$$

The two inequalities complete the proof of the Lemma.

3.3. An analytic lemma

We continue to use the notation and results of the two preceding sections.

Assume that there is an element $p \varepsilon U^2$ such that

1) $p \epsilon E(1)^2 - E(1)$.

As already remarked such an element surely exists if U does not contain a nontrivial connected group. There is a subsequence $I' \subset I$ of elements p_n, $n \epsilon I'$, such that

2) $\lim p_n = p$

and

3) $p_n = x_{n,1}x_{n,2} \ldots x_{n,2k(n)}, \quad x_{n,i} \epsilon A_n$.

Since p is not in $E(1)$ it follows that

$$E(1/2) \cap pE(1/2) = \Phi.$$

We shall now determine a compact neighborhood $X(e) \subset U$ to play the role of the arbitrary X of the preceding section. There is a compact $X(e)$ such that

4) $E(1/2)X \cap pE(1/2)X = \Phi$

and also

5) $A_n^{[k(n)/2]} X \cap p_n A_n^{[k(n)/2]} X = \Phi$

for all but a finite number of the integers n in I'.

We now select a function θ_0 as described in the preceding section, with carrier in the set X just determined. Then precisely as before, we define the functions θ_n.

Now let us choose a left invariant integral on G; let φ_u denote a continuous non-negative function with compact carrier which is equal to 1 on U^2 and let the integral on G be normalized so that

6) $\int \varphi_u = 1.$

We recall the notation

7) $(f, g) = \int fg.$

We shall deal exclusively with continuous functions having compact carriers, and thus the integrals which we indicate will always exist. For θ_0,

8) $\int \theta_0 \leq 1,$

since $\theta_0 \leq \varphi_u$. Additivity of the integral shows that

$$(f, \Sigma g_i) = \Sigma (f, g_i)$$

and invariance of the integral that

$$(af, g) = (f, a^{-1}g).$$

It follows from properties of the integral that

$$| (f, g) | \leq (| f |, | g |).$$

Since X contains the carrier of θ_0 and $A_n^{[k(n)/2]}X$ the carrier of θ_n, it follows from 5) that the carriers of θ_0 and $p_n \theta_n$ have no common point for almost all n. Therefore for large n

$$(p_n \theta_n, \theta_0) = 0,$$

and

9) $(\theta_n - p_n\theta_n, \theta_0) = (\theta_n, \theta_0) \geq (\theta_0, \theta_0) = \gamma > 0.$

Let some of the factors of p_n be singled out as follows, $y_{n,1} = e$ and

$$y_{n,i} = x_{n,1} \ldots x_{n,i-1}, \quad i = 2, \ldots, 2k(n).$$

Then

$$(\theta_n - p_n\theta_n, \theta_0) = \Sigma_{i=1}^{2k(n)}(y_{n,i}(e - x_{n,i})\theta_n, \theta_0).$$

Hence for some fixed i^* depending on n, in view of 9)

10) $k(n) \mid (y_{n,i^*}(e - x_{n,i^*})\theta_n, \theta_0) \mid \geq \gamma/2.$

We now write y_n for y_{n,i^*} and x_n for x_{n,i^*} and express 10) as follows:

A) $k(n) \mid ((e - x_n)\theta_n, y_n^{-1}\theta_0) \mid \geq \gamma/2.$

This formula gives an order of magnitude of the elements x_n regarded as operators on the functions θ_n. We seek next an order of magnitude of $x_n^{k(n)}$, and shall find it through the following identity:

$$((e - x_n^{k(n)})\theta_n, y_n^{-1}\theta_0) = (\Sigma_{i=0}^{k(n)-1}x_n^i(e - x_n)\theta_n, y_n^{-1}\theta_0)$$
$$= (k(n)(e - x_n)\theta_n, 1/k(n) \Sigma_{i=0}^{k(n)-1}x_n^{-i}y_n^{-1}\theta_0).$$

This leads directly to

B) $((e - x_n^{k(n)})\theta_n, y_n^{-1}\theta_0) = (k(n)(e - x_n)\theta_n, y_n^{-1}\theta_0)$
$+ (k(n)(e - x_n)\theta_n, 1/k(n) \Sigma_{i=0}^{k(n)-1}(x_n^{-i} - e)y_n^{-1}\theta_0).$

The first term on the right side of B) is always $\geq \gamma/2$; it follows that the other two terms cannot each be less in absolute value than $\gamma/4$.

In the last term of B) the element x_n belongs to A_n and it follows from 3.2.3 that for all x and for all n

$$k(n) \mid (e - x_n)\theta_n(x) \mid \leq k(n)\Delta_n(x_n).$$

From the definition of Δ_n this gives

C) $k(n) \mid (e - x_n)\theta_n(x) \mid \leq 2, \quad n \varepsilon I'.$

LEMMA. *Let G be locally compact, first countable, and let the subsets A_n described in 3.1.1 generate $E(1)$ and $E(1/2)$ and let $E(1)^2 \neq E(1)$. Then there exists a neighborhood $W(e)$ and for an infinite sequence $I' \subset I$ there exists $x_n \varepsilon A_n$, $n \varepsilon I'$, such that for some integer $j(n) \leq k(n)$*

$$x_n^{j(n)} \text{ is not in } W.$$

The sequence I' is determined as above so that the formulas A, B, C are valid. Now we choose W as the symmetric neighborhood V^* of 3.2.2 corresponding to $\epsilon = \gamma/8$ $(\gamma = (\theta_0, \theta_0)$, as above). Then for any a, $b \varepsilon U$ and $n = 0, 1, \ldots$,

$$|\theta_n(ax) - \theta_n(bx)| \leq \gamma/8$$

provided that $a^{-1}b$ is in W and x is confined to U^2.

Suppose now that there is an integer $n \varepsilon I'$ such that

$$x_n^i \varepsilon W, \ i = 1, 2, \ldots, k(n).$$

Then by the choice of W, taking $a = e$ and $b^{-1} = x_n^{k(n)}$,

$$|(e - x_n^{k(n)})\theta_n(x)| \leq \gamma/8.$$

By the monotonic property of the integral, it follows that the left hand side of B) is not greater in absolute value than

$$\gamma/8 \, |\textstyle\int y_n^{-1}\theta_0|$$

Using the left invariance of the integral and relation 8) above, we get

$$\gamma/8 \, |\textstyle\int y_n^{-1}\theta_0| = \gamma/8 \, |\textstyle\int\theta_0| \leq \gamma/8.$$

In order to estimate the second term on the right of B, fix a value of $i \leq k(n)$ and set $a = y_n x_n^i$ and $b = y_n$. Then $a^{-1}b = x_n^{-i} \varepsilon W$. Therefore,

$$|(x_n^{-i} - e)y_n^{-1}\theta_0(x)| = |\theta_0(y_n x_n^i x) - \theta_0(y_n x)| \leq \gamma/8$$

by the choice of W. We have to add $k(n)$ terms of this kind, for $i = 1, \ldots k(n)$, and divide by $k(n)$. The absolute value of the sum is bounded by $\gamma/8$. By C), the carrier being in U^2,

$$|k(n)(e - x_n)\theta_n(x)| \leq 2\varphi_u$$

independently of x, it follows that the absolute value of the second term on the right is at most $\gamma/4$.

As already remarked, the absolute value of the middle term of B) exceeds $\gamma/2$ and this gives a contradiction. The Lemma is proved.

3.3.2.   As a first application of this Lemma we complete the general proposition which we looked at in 2.22. We shall show that *every locally compact connected metric (non-trivial) group G contains a compact connected group or else it contains an abelian connected group.* By what has already been proved in 2.22, it follows that G contains *a one-parameter subgroup* (non-trivial).

Let U be a compact neighborhood of e and let A_n, $n = 1, 2, \ldots,$ be a sequence of compact neighborhoods of e converging to e. Since G is connected, G is generated by every neighborhood of e and it follows that each A_n has some power which is not contained in U. Therefore we can pick an appropriate subsequence of these sets to be a generating sequence in the sense of the preceding sections. Now there are two cases.

First, $E(1) = E(1)^2$. In this case $E(1)$ is a compact connected group which is non-trivial.

If $E(1) \neq E(1)^2$, then the preceding Lemma applies, and it guarantees us a neighborhood V and a sequence of elements x_n, $n = 1, 2, \ldots,$ converging to e such that each x_n has some power that is not contained in V. In this case the sequence of elements x_n will give rise to an abelian connected group. In the light of the foregoing sections, we see this most easily by defining a new generating sequence $B_1, B_2, \ldots,$ in relation to V.

Thus, let $B_n = e \cup x_n \cup x_n^{-1}$. Now the B_n's generate a set $E'(1)$ which meets the boundary of V and is compact connected and *abelian*; that is, every pair of elements of $E'(1)$ commute. *For* if $i(n)$ and $j(n)$ are sequences of integers such that $x_n^{i(n)}$ converges to some y and $x_n^{j(n)}$ converges to some element z, then $yz = zy$. The set $E'(1)$ generates an abelian connected group; and this concludes the proof.

3.4. Groups without small subgroups

In the remainder of this chapter we assume that G is locally compact, first countable, with a compact symmetric $U(e)$ such that $\{e\}$ is the only subgroup of G entirely contained in U. Most of the Lemmas can also be proved by essentially the same arguments for the case of a locally compact *connected* local group but we shall not complicate the statement and proofs of the Lemmas by inserting the necessary qualifications. The case where G is a connected locally euclidean local group may be one the interested reader would like to keep in mind.

LEMMA. *Let G be a locally compact group and let $U(e)$ be compact, symmetric, and let U contain no subgroup except e. Let K be a compact subset of U, e not in K. There is an integer $h \geq 2$ such that for $x \in K$ there is an $m(x)$, $1 < m \leq h$, such that x^m is not in U.*

For any x in G, the powers x^n form a group. Therefore if $x \in U$, $x \neq e$, there is a least positive integer $m = m(x)$ such that

$$x^m \text{ is not in } U.$$

Then there is a neighborhood of x^m which does not meet U and by the continuity of multiplication there is a $W(e)$ such that if $z \in Wx$, then also z^m is not in U. Thus $m(z) \leq m(x)$ and m is an upper semi-continuous function and has a maximum on the compact set K. This maximum is denoted by h. For every $x \in K$, $m(x) \leq h$, and the Lemma is proved.

3.4.1. COROLLARY. *Given a neighborhood $W_0(e)$, $W_0 \subset U$ above, there is a number r_0 in $(0, 1)$ depending on W_0 such that if x and n satisfy $x^i \in U$, $i \leq n$, then $x^j \in W_0$, $j \leq r_0 n$.*

By the preceding Lemma determine h for the set $K = U - W_0$, and let $r_0 = 1/2h$. If for some k, x^k is not in W_0, then x^{km} is not in U for some $m \leq h$. Then

$$km \geq n,$$
$$k \geq n/m \geq n/h > r_0 n$$

which was to be proved.

A locally compact G contains an open subgroup G' (2.3.1, 2.6) such that in every $U(e) \subset G'$ there is a compact subgroup H with G'/H separable. In the present case when U has no non-trivial subgroup, $H = e$ and G' is separable so that G is locally separable. Hence we may always choose in G a compact separable $W(e)$ such that W, or W^2, contains no non-trivial subgroup.

3.5. Uniqueness of square roots

The following Lemma was proved by Gleason [3] and by Kuranishi [2]. Compare also Chevalley [3].

LEMMA. *Let G be a locally compact group without small subgroups. There exists a neighborhood U of e such that if x and y are in U and $x^2 = y^2$ then $x = y$.*

Let W be a symmetric compact separable (see last paragraph of 3.4) neighborhood of e containing no non-trivial subgroup. Suppose that

$$x^2 = y^2$$

for some x and y in W and let

$$a = x^{-1}y.$$

Then

$$x^{-1}ax = x^{-2}yx = y^{-1}x = a^{-1}$$

and for every integer m

$$x^{-1}a^m x = (x^{-1}ax)^m = a^{-m}.$$

Assume now that the Lemma is false. There must then exist a sequence of pairs of elements x_n and y_n with

$$\lim x_n = \lim y_n = e$$

and

$$x_n^2 = y_n^2,$$

but with

$$a_n = x_n^{-1}y_n \neq e.$$

There is also a sequence of integers $m(n)$ for some sequence

$n \in I' \subset I$ such that

$$a_n^{m(n)} \in W, \ a_n^{m(n)+1} \notin W$$

and

$$\lim a_n^{m(n)} = b$$

exists. Since $\lim a_n = e$, it follows that

$$\lim a_n^{m(n)+1} = b$$

and therefore

$$b \in \text{boundary } W, \ b \neq e.$$

But

$$x_n^{-1} a_n^{m(n)} x_n = a_n^{-m(n)}$$

and by taking limits we get

$$b = b^{-1}, \ b^2 = e$$

and the set consisting of e and b is a non-trivial group in W. This is a contradiction and proves the Lemma.

3.5.1. Let U denote a compact symmetric separable neighborhood of e *such that U contains no non-trivial group and such that if $x, y \in U$ and $x^2 = y^2$ then $x = y$.*

COROLLARY. *If $x(t)$ and $y(s)$, $s, t \in [-1, 1]$ are one-parameter local subgroups of U:*

$$x(t), \ y(s) \in U, \ s, \ t \in [-1, 1]$$

and if $x(t_0) = y(s_0)$ for some $s_0, t_0 \in [-1, 1]$, then $x(rt_0) = y(rs_0)$ for $r \varepsilon [-1, 1]$.

For the proof, observe that

$$x(t_0) = x(t_0/2)^2 = y(s_0/2)^2 = y(s_0)$$

and consequently

$$x(t_0/2) = y(s_0/2).$$

Inductively

$$x(t_0/2^h) = y(s_0/2^h)$$

for every h and $x(kt_0/2^h) = y(ks_0/2^h)$ for every integer $k \leq 2^h$. Therefore by continuity

$$x(rt_0) = y(rs_0) \qquad |r| \leq 1$$

proving the Lemma.

We have introduced the Corollary at this point to indicate that the one-parameter groups in U will "radiate from the identity," suggestive of their behavior in Lie groups. However we shall not prove until nearly the end of the major theorem that all elements in some neighborhood of e are on one-parameter groups and have square roots.

3.6. The family of one-parameter local groups

Let U be a compact, symmetric neighborhood with unique square roots and with no subgroup except e.

LEMMA 1. *Suppose that* $x_n \varepsilon U$, $n \in I'$, *and suppose that an increasing sequence of integers* $h(n)$ *is given such that* $x_n^i \varepsilon U$ *for* $i \leq h(n)$. *If* V *is an open neighborhood of* e, *there exists* $r_0 > 0$ *such that*

$$x_n^i \epsilon V$$

for $i \leq [r_0 h(n)]$.

This follows from 3.4.1.

COROLLARY 1. *For the* x_n *as above,* $\lim x_n = e$. *Also,* $\lim x_n^{j(n)} = e$ *for every sequence of integers* $j(n)$ *for which* $\lim j(n)/h(n) = 0$.

LEMMA 2. *With the same hypotheses as in Lemma 1 except that moreover*

$$\lim x_n^{h(n)} = x$$

exists, it follows that for every r, $0 \leq r \leq 1$

1) $\lim x_n^{[rh(n)]}$ *exists.*

If $\lim x_n^{[rh(n)]}$ *is denoted by* $x(r)$ *and if* r_1, r_2, $r_1 + r_2$ *are in* $[0, 1]$ *then*

2) $x(r_1)x(r_2) = x(r_1 + r_2)$.

There exists a subsequence $I^* \subset I'$ such that for every rational r in $[0, 1]$

$$x(r) = \lim x_n^{[rh(n)]}, \quad n \, \epsilon \, I^*$$

exists (for $r = 0$, set $x_n^0 = e$). For r_1, r_2, $r_1 + r_2$ rationals in $[0, 1]$ it follows easily from $\lim x_n = e$ that

3) $x(r_1 + r_2) = x(r_1)x(r_2).$

3.6.1.   To prove 1) we shall now employ a type of argument that will be useful to us in several lemmas: we are given a sequence about which we know that it and every subsequence of it has *convergent subsequences*. We then show that the *limit* of such a convergent subsequence can be determined independently of the choice of subsequence. It follows that the limit is unique and that the original sequence converges.

3.6.2.   Now, pick a definite r in $[0, 1]$ and suppose that $z = \lim x_n^{[rh(n)]}$ exists for n in an appropriate subsequence of I^*. Let $W \subset U$ be an arbitrary neighborhood (compact, symmetric) of e and choose $r_0 > 0$ in accordance with 3.4.1. Let r' denote any rational in the interval $r < r' < r + r_0$. Define integers $j(n)$ by the equation

$$x_n^{[r'h(n)]} = x_n^{[rh(n)]} x_n^{j(n)}.$$

It follows (in the limit on n) from the choice of r_0 that

$$x(r') \, \epsilon \, zW.$$

Now, since W is arbitrary we see that if r_m, $m \, \epsilon \, I$, is any sequence of rationals for which $\lim r_m = r$, then $\lim x(r_m) = z$. Thus $x(r)$ and $\lim x(r_m)$ are the same for these subsequences of I^*. Therefore, finally, $\lim x_n^{[rh(n)]}$ exists for $n \, \epsilon \, I^*$.

The argument just given shows also that $x(r_m)$ converges to $x(r)$ when r_m converges to r. Since 2) holds for rationals (by 3) above), 2) is valid for reals by a passage to the limit.

Now we use the type of argument mentioned above (3.6.1) to see that 1) and 2) hold for limits defined over the original sequence I'. This follows from the fact that the one-parameter local semigroup which we have just got is uniquely determined by the element $x(1)$, $x(1) = \lim x_n^{h(n)}$. This concludes the proof of the Lemma.

3.6.3. LEMMA 3. *With the same hypotheses as above, given $r \epsilon [0, 1]$ and a sequence of integers $i(n) \leqq h(n)$ with $\lim i(n)/h(n) = r$, then*

$$\lim x_n^{i(n)} = x(r).$$

For $r = 0$ this follows from Lemma 1. In general, $i(n) = [rh(n)] + j(n)$ and $\lim j(n)/h(n) = 0$. Then $\lim x_n^{j(n)} = e$ and

$$\lim x_n^{i(n)} = \lim x_n^{[rh(n)]} \cdot \lim x_n^{j(n)} = x(r).$$

COROLLARY 2. *With the same hypothesis as in Lemma 2 if a neighborhood V of e is given and a sequence of integers $h'(n) \leqq h(n)$ then for almost all n*

$$x_n^{-h'(n)} x(h'(n)/h(n)) \epsilon V.$$

The corollary follows at once for any sequence $h'(n)$ for which the ratio $h'(n)/h(n)$ converges. Therefore it holds for all sequences $h'(n)$.

3.6.4. LEMMA 4. *Let a neighborhood $V(e)$, $V \subset U$ be given. There is a neighborhood $W(e)$ such that for every sufficiently large integer k*

    1) $x, x^2, \ldots, x^k \epsilon U$, $y, y^2, \ldots, y^k \epsilon U$

and

    2) $x^k y^{-k} \epsilon W$

imply that

$$x^i y^{-i} \epsilon V, \quad i \leqq k.$$

If this Lemma is false there exist sequences of elements x_n and y_n and there exists a sequence of increasing integers $h(n)$ such that

$$x_n, \ldots, x_n^{h(n)} \epsilon U$$
$$y_n, \ldots, y_n^{h(n)} \epsilon U$$
$$\lim x_n^{h(n)} y_n^{-h(n)} = e;$$

but for some $i(n) \leqq h(n)$

$$x_n^{i(n)} y_n^{-i(n)} \notin V.$$

Since the sequence $h(n)$ increases without limit, the sequences x_n and y_n must converge to e (Corollary 1). We suppose that both sequences have been "thinned" so that, also,

$x_n^{h(n)}$ converges to an element $x(1)$, and $y_n^{h(n)}$ to an element $y(1)$.

It is clear that $x(1) = y(1)$ and therefore (3.5.1) $x(r) = y(r)$, $r \in [0, 1]$. Choose $V_1(e)$ so that $V_1^2 \subset V$.

For all sufficiently large n (letting $i = i(n)$, $h = h(n)$)

$$x_n^i x(-i/h) \in V_1$$

and

$$x(i/h) y_n^{-i} \in V_1$$

and the product is in $V_1^2 \subset V$. This contradiction proves the Lemma.

3.6.5.   LEMMA 5.   *Given a neighborhood $W(e)$ there is an r_0, $0 < r_0 \leqq 1$ such that if $x(r)$ is a one-parameter local group belonging to U for $|r| \leqq 1$, then*

$$x(s) \in W \text{ if } |s| \leqq r_0.$$

This is a consequence of Lemma 1 and Corollary 2.

3.6.6.   LEMMA 6.   *If $\lim x_n^{h(n)} = x = e$, $x_n^i \in U$, $i \leqq h(n)$, then $x(r) \equiv e$ and*

$$\lim x_n^{i(n)} = e, \; i(n) \leqq h(n).$$

This follows from 3.5.1 because $e(r) \equiv e$ is a one-parameter semigroup and so is $x(r) = \lim x_n^{[rh(n)]}$; and by assumption $x(1) = e(1)$.

The following lemma gives the beginnings of a view of the *family* of (local) one-parameter semigroups which exist in U.

3.6.7.   LEMMA 7.   *Suppose that for each $n \varepsilon I$, $z_n(t) \subset U$ is a one-parameter semigroup, $t \varepsilon [0, 1]$, and suppose that $z_n(1)$ converges to an element z_0. Then z_0 is on a semigroup $z_0(t)$ and for every t, $z_n(t)$ converges to $z_0(t)$. The convergence is uniform.*

First, let us consider only dyadic rational values of the parameter t. Every convergent subsequence of $z_n(1/2)$ converges to the unique square root of z_0. Then $z_n(1/2)$ converges, and we may denote the *limit* by $z_0(1/2)$. By induction, it follows that $z_n(k/2^m)$ converges to an element which we now denote by $z_0(k/2^m)$, k and m integers, $k/2^m \, \varepsilon \, [0, 1]$. By Lemma 2, the sequence $z_0(1/2^m)$ generates a one-parameter semigroup $z_0(t)$ with $z_0(1) = z_0$.

Now let symmetric $U_1(e)$, $V(e)$ be given with $V^3 \subset U_1 \subset U$. Let V determine a $W(e)$ in accordance with Lemma 4 and let V determine $r_0 > 0$ in accordance with Lemma 5. By hypothesis if n is sufficiently large, say $n \geqq N$, then

$$z_n(1) \, \varepsilon \, W z_0,$$

and therefore, for every pair of integers k and h, $k \leqq h$,

$$z_n(k/h) \, \varepsilon \, V z_0(k/h).$$

If t, t', $t' - t \, \varepsilon \, [0, 1]$ and $t' - t < r_0$, then for all $n = 0, 1, \ldots$, by the choice of r_0:

$$z_n(t') \, (z_n(t))^{-1} = z_n(t' - t) \, \varepsilon \, V.$$

Fix t'. Now choose $h = 2^m > 1/r_0$ and choose an integer k so that for $t = k/h$, $t' - t < r_0$. Finally, from $z_n(t') \, \varepsilon \, z_n(t)V$, $z_n(t) \, \varepsilon \, z_0(t)V$, $z_0(t) \, \varepsilon \, z_0(t')V$ (because V is symmetric) we arrive at

$$z_n(t') \, \varepsilon \, z_0(t')V^3 \subset z_0(t')U_1, \quad n > N.$$

In this formula, U_1 is an arbitrary preassigned neighborhood of e, and $t' \, \varepsilon \, [0, 1]$; the integer N depends only on U_1. This shows that the convergence of $z_n(t)$ to $z(t)$ is uniform.

Incidentally we have shown that the set of elements of U which lie on one-parameter semigroups is compact.

3.6.8. So far we have spoken mainly of semigroups. If $x(t)$ is a local semigroup in a symmetric neighborhood U, then the set of inverses is also a semigroup, and the union of these two semigroups (using an appropriate parameter $t \, \varepsilon \, [-1, 1]$) is a *local one-parameter group* in U:

$$x(t + t') = x(t)x(t'), \quad \text{if } t, \, t', \, t + t' \, \varepsilon \, [-1, 1].$$

The details are quite elementary and will be omitted except for the remark that the appropriate parameter to use for the local group is one which extends one of the semigroups by giving the negative parameter value to each inverse element. The reader will bear in mind that the elements of a one-parameter semigroup commute (by 2) of Lemma 2); this is also true for the local group.

3.7. The set of n-th roots

Let U_0 be a separable compact symmetric neighborhood (p. 120) of e such that U_0^8 has e as its only subgroup. The reader will understand that the use of one or another power of U_0 is designed to keep certain group products within a region of "no subgroups".

To make available the machinery which was set up in the beginning of the chapter we now choose as generating sets A_n (3.1), $n \varepsilon I$, the set of all elements all of whose powers up to the n-th belong to U_0

$$A_n = \{x; \ x^i \varepsilon U_0, \ i = 1, \ldots, n\} \quad n \varepsilon I.$$

Each A_n is a compact symmetric subset of U_0 containing e and $\lim A_n = e$ by Lemma 3.4. We discard the trivial case where the group is discrete and the identity is an isolated element of U_0. It follows that A_n is never equal to e and that for each n there is a *least* integer $k(n)$ such that

$$A_n^{k(n)} \subset U_0, \ A_n^{k(n)+1} \not\subset U_0.$$

We may suppose now that the construction of sections $3.1-3.3$ has been repeated with this choice of generating sets. Then $E(1/2)$ and $E(1)$ are defined as before

$$E(1/2)^2 \subset E(1) \subset U_0$$

and now because there are no subgroups in U_0

$$E(1)^2 \neq E(1).$$

Select any $p \varepsilon E(1)^2 - E(1)$ and choose a compact symmetric neighborhood X of e restricted as before by

1) $\quad pA_n^{[k(n)/2]} X \cap A_n^{[k(n)/2]} X = \Phi$

for sufficiently large $n \, \varepsilon \, I'$ (see **3.3**) and also by the condition $X^2 \subset U_0$, so that

2) $X^2 U_0 \subset U_0^2.$

Define the function θ_0 as before, with carrier in X and define the functions $\varDelta_n$ and θ_n for *all* $n \, \varepsilon \, I$.

LEMMA.   *If $q \notin U_0^2$ then the inner product $(q\theta_n, \, \theta_0)$ vanishes identically.   This holds for $n \, \varepsilon \, I$.*

The carrier of θ_0 is X. The carrier of θ_n (see 3.2) is contained in $A_n^{[k(n)/2]} X \subset U_0 X$. Therefore the carrier of $q\theta_n$ is $q U_0 X$. All that we need to prove is that $q U_0 X \cap X = \varPhi$. Suppose there could exist $x, \, y \, \varepsilon \, X$ and $u \, \varepsilon \, U_0$ such that $qux = y$. It would follow from the symmetry of X and of U_0 that $q \, \varepsilon \, X^2 U_0 \subset U_0^2$. This is contrary to hypothesis, and the Lemma is proved.

3.7.1.   In defining the sets $E(1)$, $E(1/2)$ we use a *subsequence* of the sequence of sets

$$A_1, A_2, \ldots, A_n, \ldots$$

and in proving the Lemma of 3.1.1 we restrict the subsequence slightly more. Therefore in order to apply *all* of the results of 3.1, 3.2, 3.3 we need to restrict ourselves to sets A_n, $n \, \varepsilon \, I'$ and n large. However after we have proved the following Lemma we shall be able to free ourselves of this particular restriction.

LEMMA.   *The integers $k(n)$ associated with the generating sets A_n, $n \, \varepsilon \, I'$*

$$A_n = \{x; \, x, \, x^2, \ldots, x^n \, \varepsilon \, U_0\}, \, n \, \varepsilon \, I',$$

satisfy a relation

$$n < k_0 k(n)$$

for a constant (integer) k_0.

If we apply Lemma 3.3 we see that there is a $V_0(e)$ and in each A_n an element x_n such that for an $i(n) \leq k(n)$ depending on x_n, $n \, \varepsilon \, I'$,

$$x_n^{i(n)} \notin V_0.$$

There is an integer k_0 associated with this V_0 by 3.4 such that for some $j(n) \leq k_0 k(n)$ depending on x_n,

$$x_n^{j(n)} \notin U_0.$$

But then $j(n) > n$ by the definition of A_n and $n < j(n) \leq k_0 k(n)$. This concludes the proof.

3.7.2. The sets A_n are defined for all n and so are the functions $k(n)$. This is all we need at the moment to free the preceding Lemma from dependence on a particular sequence I'.

COROLLARY 1. *There is an integer k_0 such that for all $n \varepsilon I$,*

1) $n < k_0 k(n)$.

If the Corollary were not true then there would be an infinite sequence $I'' \subset I$ such that the ratio $n/k(n)$, $n \varepsilon I''$, increased without limit as n increased. We now apply all the preceding construction of this section to the sets A_n, $n \varepsilon I''$. Finally we prove the Lemma, above, for the numbers n and $k(n)$, *in the present case* that $n \varepsilon I''$. This gives us a constant which is an upper bound to the ratios $n/k(n)$, $n \varepsilon I''$. The contradiction proves the Corollary.

We *now* denote by k_0 the constant which has been proved to exist such that 1) holds *for all n*.

COROLLARY 2. *If x_n, $y_n \varepsilon A_n$, then*

$$(x_n y_n)^i \varepsilon U_0^{2k_0}, \quad i \leq n.$$

It is clear that $(x_n y_n)^i \varepsilon A_n^{2n}$, $i \leq n$. But

$$A_n^{2n} \subset A_n^{2k_0 k(n)} \subset (A_n^{k(n)})^{2k_0} \subset U_0^{2k_0}$$

and this proves the corollary.

3.7.3. It follows at once that *if x_n and y_n are sequences, $n \varepsilon I^*$, $(x_n y_n)^n$ is in a compact set (2.1.1) and has a convergent subsequence.* This will be of especial interest for the case $(x(1/n)y(1/n))^n$, where $x(t)$, $y(t)$ are one-parameter local groups, $t \varepsilon [-1, 1]$, contained in U_0. To investigate this situation we shall again regard the elements of G as operators on continuous functions with compact carriers. We use the left invariant integral on G normalized so that $\int \varphi_U = 1$, where φ_U is 1 on U_0^3, has a compact carrier, and has all its values in $[0, 1]$.

3.7.4.   There is another formulation of the results of 3.7.2 which will be particularly convenient when we deal with one-parameter local groups (in 3.9).

   LEMMA. *For every $n \varepsilon I$, if $x \varepsilon A_{k_0 n}$ and $i \leq n$, then $x^i \varepsilon A^n_{k_0 n} \subset U_0$.*
The function k is defined for every n and therefore for $k_0 n$; then

$$k_0 n < k_0 k(k_0 n)$$

by 3.7.2 and, equivalently,

$$n < k(k_0 n).$$

Therefore, finally, with $i \leq n$

$$x^i \varepsilon A^{k(k_0 n)}_{k_0 n} \subset U_0$$

by the definition of the function k.   This completes the proof.

3.7.5.   The only way in which we shall use the Lemma of 3.3 in this Chapter is through the Lemma of 3.7.4 which is valid for all n.   The functions Δ_n and θ_n are defined for all n (in 3.2) and *the important Lemmas in 3.2.3 and 3.7 are valid for all n.*   The functions θ_n depend on θ_0 as well as on the functions Δ_n; the function θ_0 was required to vanish outside of a neighborhood X whose construction depended on a subsequence of I but this fact will not enter into our use of the functions θ_n and Δ_n.

   Therefore we do not need to be tied to the sequence I' used in 3.1, and hereafter the symbol I' will denote a subsequence as needed.

3.8. A key lemma

   Let k_0 be the integer such that $n < k_0 k(n)$ for all n (3.7.2). In the following we shall often use the exponent $[n/k_0]$ and for convenience we define and use

$$n^* = [n/k_0].$$

Then for all n, by 3.7.2,

$$n^* < k(n).$$

We shall consider the two quantities $x_n^{n^*} y_n^{n^*}$ and $(x_n y_n)^{n^*}$, for x_n, $y_n \epsilon A_n^4$, $n \epsilon I^* \subset I$, and prove in the next two lemmas that if either of these converges to e then so does the other.

The elements x_n, y_n are taken in A_n^4 instead of in A_n to allow for later application. We use the exponent n^*, rather than n, to be certain that $(x_n y_n)^{n^*}$ and $x_n^{n^*} y_n^{n^*}$ are in U_0^8 where we know that there are no subgroups except e. Thus for x_n, $y_n \epsilon A_n^4$ and $i \leqq n^*$, we have

$$(x_n y_n)^i \epsilon A_n^{8n^*} \subset A_n^{8k(n)} \subset U_0^8,$$
$$x_n^i y_n^i \epsilon A_n^{4n^*} A_n^{4n^*} \subset U_0^8.$$

LEMMA. *With notation as in 3.7 let some open $V(e) \subset U_0$ be given. There is a $W(e)$ such that for every sequence of pairs x_n, $y_n \epsilon A_n^4$, $n \epsilon I^*$,*

$$x_n^{n^*} y_n^{n^*} \epsilon W$$

implies that

$$(x_n y_n)^{n^*} \epsilon V$$

for almost all n.

Let h be the integer associated with $U_0 - V$ and with U_0^2 by 3.4; then if $z \notin V$, z^i is not in U_0^2 for some $i \leqq h$. Let k_0 be the constant (3.7.4) such that $n < k_0 k(n)$ and choose $\epsilon_0 > 0$ by

$$10h \, \epsilon_0 = (\theta_0, \, \theta_0)$$

where θ_0 is defined as before (see top p. 128).

Pick a symmetric neighborhood W_1 of e such that, independently of $y \epsilon U_0$ (it is understood that $W_1 \subset U_0$)

$$| \theta_0(ay) - \theta_0(by) | < \varepsilon_0/8$$

if $a^{-1}b$ or $ab^{-1} \epsilon W_1$; this was shown to be possible in section 3.2.2.

If $x \epsilon A_n^4$, then $x^i \epsilon U_0^8$ for $i \leqq n^*$. Therefore we can choose $r_0 \epsilon (0, 1)$ so that if $x \epsilon A_n^4$ then $x^j \epsilon W_1$ for $j \leqq [r_0 n^*]$. This was proved to be possible independently of n in section 3.4.1.

Next, by 3.2.3 and an appropriate change of variables there is a symmetric neighborhood $W_2(e) \subset W_1$ such that

$$|(a - b)\theta_n(x)| < r_0 \,\epsilon_0/8$$

if $a^{-1}b$ or $ab^{-1} \,\varepsilon\, W_2$ (see bottom p. 112 for notation).

Finally the desired $W \subset W_2$ is defined so that if $x_n^{n^*} y_n^{n^*} \,\epsilon\, W$ then $x_n^i y_n^i \,\epsilon\, W_2$ for all $i \leq n^*$. This is possible by Lemma 4 in 3.6.4 (to apply Lemma 4, replace y of that Lemma by y^{-1}).

For all sufficiently large n

$$A_n^4 \subset W.$$

Assume now for some n for which $n^* > 1/r_0$ and $A_n^4 \subset W$, that

$$x_n^{n^*} y_n^{n^*} \,\epsilon\, W$$

but

$$(x_n y_n)^{n^*} \,\notin\, V.$$

We shall obtain a contradiction.

Since $(x_n y_n)^{n^*} \,\notin\, V$, it follows that there is an integer $m \leq h$ such that $(x_n y_n)^{mn^*} \,\notin\, U_0^2$. Then using $(x_n y_n)^{mn^*}$ as the element q of the Lemma of 3.7 we see that

$$((x_n y_n)^{mn^*}\theta_n, \; \theta_0) = 0.$$

Now

1) $10h\,\epsilon_0 = (\theta_0, \,\theta_0) \leq (\theta_n, \,\theta_0) = ((e - (x_n y_n)^{mn^*})\theta_n, \,\theta_0).$

Expand the right hand term by the device used in Lemma 3.3 to obtain

$$m\big(n^*(e - x_n y_n)\theta_n, \; (1/mn^*) \textstyle\sum_{i=0}^{mn^*-1} (x_n y_n)^{-i}\theta_0\big).$$

Introduce the abbreviation

2) $\varphi = (1/mn^*) \sum_0^{mn^*-1}(x_n y_n)^{-i}\theta_0,$

and, for later use, notice that because of the invariance of the integral

2.1) $\int\varphi = \int\theta_0 \leq 1.$

Make a further expansion of 1) getting

3) $10h\,\epsilon_0 \leq m\big(n^*(e-x_n)(y_n-e)\theta_n + n^*(e - x_n)\theta_n + $
$$n^*(e - y_n)\theta_n, \; \varphi\big).$$

Consider the following term in 3):

$$(n^*(e - x_n)(y_n - e)\theta_n, \varphi) = (n^*(y_n - e)\theta_n, (e - x_n^{-1})\varphi).$$

Since $y_n \epsilon A_n^4$, and so by 3.2.3

$$| (y_n - e)\theta_n(x) | \leq \Delta_n(y_n) \leq 8/k(n),$$

it follows that

$$| n^*(y_n - e)\theta_n(x) | < 8.$$

The $(i + 1)$th summand in the function $(e - x_n^{-1})\varphi$ is

3.1) $\theta_0((x_n y_n)^i x) - \theta_0((x_n y_n)^i x_n x).$

Let $(x_n y_n)^i = a$ and let $(x_n y_n)^i x_n = b$. Then $a^{-1}b = x_n \epsilon A_n^4 \subset W \subset W_1$. Then the absolute value of 3.1) is less than $\epsilon_0/8$ by the choice of W_1. There are mn^* summands in 2), and the coefficient $(1/mn^*)$ so that

$$| (e - x_n^{-1})\varphi(x) | < \epsilon_0/8.$$

Finally from the monotonic property of the integral and from the way the integral was normalized

4) $| m(n^*(e - x_n)(y_n - e)\theta_n, \varphi) | < h\epsilon_0.$

Next consider

5) $([r_0 n^*](e - x_n)\theta_n - (e - x_n^{[r_0 n^*]})\theta_n, \varphi) =$

$$= (\Sigma_{i=0}^{[r_0 n^*]-1}((e - x_n) - x_n^i(e - x_n))\theta_n, \varphi)$$

$$= ([r_0 n^*](e - x_n)\theta_n, 1/[r_0 n^*] \Sigma_{j=0}^{[r_0 n^*]-1}(e - x_n^{-j}) \varphi).$$

Since $j < [r_0 n^*]$, it follows from the choice of r_0 that x_n^j, and therefore also x_n^{-j}, belongs to W_1. Then by the choice of W_1, for each j

$$| (e - x_n^{-j})\varphi(x) | < \epsilon_0/8.$$

Summing on j and dividing by $[r_0 n^*]$ we get

$$1/[r_0 n^*] | \Sigma(x_n^{-j} - e) \varphi | < \epsilon_0/8.$$

Recall again that $n^* < k(n)$. Since $| [r_0 n^*](e - x_n) \theta_n(x) |$ is bounded by $[r_0 n^*]8/k(n)$, hence by $8r_0$, we get from 5),

6) $|([r_0 n^*](e - x_n)\theta_n - (e - x_n^{[r_0 n^*]})\theta_n, \varphi) | < r_0 \epsilon_0.$

By precisely similar steps

7) $\quad |([r_0n^*](e - y_n)\theta_n - (e - y_n^{[r_0n^*]})\theta_n, \varphi)| < r_0\epsilon_0.$

There remains to consider and to combine with 6) and 7) the expression

7.1) $\quad ((e - x_n^{[r_0n^*]})\theta_n + (e - y_n^{[r_0n^*]})\theta_n, \varphi) =$

$= ((e - x_n^{[r_0n^*]})\theta_n, \varphi) + ((y_n^{-[r_0n^*]} - e)\theta_n, \varphi + (y_n^{-[r_0n^*]} - e)\varphi)$

$= ((y_n^{-[r_0n^*]} - x_n^{[r_0n^*]})\theta_n, \varphi) - ((y_n^{-[r_0n^*]} - e)\theta_n, (y_n^{-[r_0n^*]} - e)\varphi).$

By the choice of W, the fact that $x_n^{n^*} y_n^{n^*} \in W$ implies

$$x_n^{[r_0n^*]} y_n^{[r_0n^*]} \in W_2$$

and, therefore, by the choice of W_2, (with $a = x_n^{[r_0n^*]}$, $b = y_n^{-[r_0n^*]}$)

$$| (y_n^{-[r_0n^*]} - x_n^{[r_0n^*]})\theta_n(x) | < r_0\epsilon_0/8.$$

Therefore since $\int | \varphi | \leq 1$ the absolute value of the first integral on the right hand end of 7.1. is bounded by $r_0 \epsilon_0/8$. Finally in the second integral, because $y_n^{-i} \varepsilon W_1$ for $i \leq [r_0n^*]$,

$$| (y_n^{-[r_0n^*]} - e)\varphi(x) | < \epsilon_0/8.$$

Since

$$\Delta_n(y_n^{-[r_0n^*]}) \leq \frac{8[r_0n^*]}{k(n)} \leq \frac{8[r_0n^*]k_0}{n} < 8r_0$$

we have

$$| (y_n^{-[r_0n^*]} - e)\theta_n(x) | < 8r_0.$$

The absolute value of the second integral (right hand end of 7.1)) is certainly less than $r_0 \epsilon_0$; and

8) $\quad |((e - x_n^{[r_0n^*]})\theta_n + (e - y_n^{[r_0n^*]})\theta_n, \varphi)| < r_0\epsilon_0 + r_0\epsilon_0/8 < 2r_0\epsilon_0.$

Combining 6), 7), and 8) we have

$$| ([r_0n^*](e - x_n)\theta_n + [r_0n^*](e - y_n)\theta_n, \varphi) | < 4r_0\epsilon_0.$$

Since

$$r_0n^* < [r_0n^*] + 1 \leq 2[r_0n^*],$$

we have

9) $|(n^*(e - x_n)\theta_n + n^*(e - y_n)\theta_n, \varphi)| < 8\epsilon_0.$

Then

10) $m\,|\,(n^*(e - x_n)\theta_n + n^*(e - y_n)\theta_n, \varphi)\,| < 8h\epsilon_0.$

Finally using 1), 4), 10) in 3) we obtain the contradiction

$$10h\,\epsilon_0 \leqq (\theta_n,\,\theta_0) < 9h\epsilon_0.$$

This concludes the proof of the Lemma.

3.8.1. Now we prove the converse.

LEMMA. *Let $I' \subset I$ be an arbitrary sequence and let $x_n,\,y_n\,\varepsilon\,A_n^4$, $n\,\varepsilon\,I'$, be such that* $\lim (x_n y_n)^{n^*} = e$. *Then* $\lim x_n^{n^*} y_n^{n^*} = e$.

By hypothesis $\lim (x_n y_n)^{n^*} = e$. Then, by 3.6.6 $\lim (x_n y_n)^{[rn^*]} = z(r)$ is a one-parameter group such that $z(r) = e$ for all r, and $\lim (x_n y_n)^{i(n)} = e$ for all $i(n) \leqq n^*$. Therefore given an arbitrary neighborhood $X(e)$,

$$(x_n y_n)^i\,\varepsilon\,X,\ i \leqq n^*$$

for all sufficiently large values of n^*.

Now let V be a fixed compact symmetric neighborhood of e such that

$$q^{-1}Vq \subset U_0$$

for all q in the compact set U_0^4 (2.4).

There is a sequence of integers $n' = n'(n)$ such that for sufficiently large n (if the Lemma is false),

1) $(x_n y_n)^j\,\varepsilon\,V,\ j \leqq n'$ but $(x_n y_n)^{n'+1} \notin V$.

Because the neighborhood X mentioned is an arbitrary neighborhood, it follows now that for the fixed neighborhood V, the integers n' satisfy

2) $\lim n^*/n' = 0,\ \ n\,\varepsilon\,I'.$

Then for all $q\,\varepsilon\,U_0^4$ and $i \leqq n'$

3) $(q^{-1}x_n y_n q)^i = q^{-1}(x_n y_n)^i q\,\varepsilon\,U_0.$

This shows that $(q^{-1}x_n y_n q)\,\varepsilon\,A_{n'}.$

Suppose now that $x_n^{n*} y_n^{n*}$ does not converge to e. There is then a $W(e)$ such that infinitely many of the $x_n^{n*} y_n^{n*}$ are not in W, and then there is a subsequence $I'' \subset I'$ and an integer h such that

$$(x_n^{n*} y_n^{n*})^h \notin U_0^2, \quad n \in I''.$$

Consider

$$0 < (\theta_0, \theta_0) \leq |(((x_n^{n*} y_n^{n*})^h - e)\theta_{n'}, \theta_0)|$$

$$= |(\sum_{i=0}^{h-1} (x_n^{n*} y_n^{n*})^i (x_n^{n*} y_n^{n*} - e)\theta_{n'}, \theta_0)|$$

$$= |((x_n^{n*} y_n^{n*} - e)\theta_{n'}, \sum_{i=0}^{h-1} (x_n^{n*} y_n^{n*})^{-i} \theta_0)|$$

$$= |((y_n^{n*} - x_n^{-n*})\theta_{n'}, x_n^{-n*} \sum_{i=0}^{h-1} (x_n^{n*} y_u^{n*})^{-i} \theta_0)|$$

$$= |\sum_{i=0}^{n*-1} ((y_n^{n*-i} x_n^{-i} - y_n^{n*-i-1} x_n^{-i-1})\theta_{n'}, \varphi_n)|$$

where

$$\varphi_n = x^{-n*} \sum_{i=0}^{h-1} (x_n^{n*} y_n^{n*})^{-i} \theta_0.$$

Continuing the expansion we get:

$$(\theta_0, \theta_0) \leq |\sum_{i=0}^{n*-1} ((x_n^i (x_n y_n) x_n^{-i} - e)\theta_{n'}, x_n^{i+1} y_n^{-n*+i+1} \varphi_n)|$$

$$\leq |n*((x_n^{i'}(x_n y_n) x_n^{-i'} - e)\theta_{n'}, \psi_n)| = \sigma_n$$

where σ_n is a number and where

$$\psi_n = x_n^{i'+1} y_n^{-n*+i'-1} \varphi_n;$$

here $i' = i'(n*)$ is a particular value of the index of summation for which the corresponding inner-product has the maximum absolute value among all the $n*$ summands.

Because of the invariance of the integral

$$\int \psi_n = \int \varphi_n = h \int \theta_0 \leq h.$$

It is clear that φ_n and ψ_n are continuous functions with compact carriers so that the inner products above are defined.

Now because $x_n^{i'} \varepsilon A_n^{4i'} \subset A_n^{4n^*} \subset U_0^4$ it follows as in 3) above that

$$x_n^{i'}(x_n y_n)x_n^{-i'} \varepsilon A_{n'}.$$

Therefore, by 3.2.3, valid for all $n \varepsilon I$ (see 3.7.5),

$$| n^*(x_n^{i'}(x_n y_n)x_n^{-i'} - e)\theta_{n'} | \leq 2n^*/k(n')$$
$$= (2n^*/n')(n'/k(n'))$$
$$< 2k_0 n^*/n',$$

by 3.7.2 valid for all $n \varepsilon I$.

Therefore, finally,

$$(\theta_0, \theta_0) \leq \sigma_n < 2hk_0 n^*/n'.$$

But because $\lim n^*/n' = 0$ this gives a contradiction for all large values of n. This completes the proof.

3.8.2. We will use the preceding results in the following particular form.

LEMMA. *Suppose we are given a_n, b_n, $c_n \varepsilon A_n^2$, $n \varepsilon I' \subset I$, and that $\lim a_n^{n^*} = a$, $\lim b_n^{n^*} = b$, $\lim c_n^{n^*} = c$ and also that $\lim(a_n b_n)^{n^*} = B$, $\lim (a_n c_n)^{n^*} = C$. Then $b = c$ if and only if $B = C$.*

First, if $b = c$, then $c^{-1}b = e$ and $\lim c_n^{-n^*} b_n^{n^*} = e$. Then by 3.8 $\lim (c_n^{-1} b_n)^{n^*} = e$; equivalently, $\lim ((c_n^{-1} a_n^{-1})(a_n b_n))^{n^*} = e$. Finally, from 3.8.1, it follows that

$$\lim (c_n^{-1} a_n^{-1})^{n^*}(a_n b_n)^{n^*} = e$$

or, equivalently, that $C^{-1}B$ is e and $B = C$.

The converse is proved by retracing the steps of this argument, using 3.8 and 3.8.1 (again in that order).

3.9. Study of $(x(1/m)y(1/m))^m$

It will be clear later (4.2) that all elements sufficiently near to e belong to local one-parameter groups in U_0. In consequence, the operation which we are going to define will apply to *all* elements near e. However we cannot show this now and for the present we confine our attention to the class of elements on local one-parameter groups contained in U_0 (except when operations performed on these elements appear to take us outside of the class).

Let $x \varepsilon U_0$ belong to a one-parameter local group contained in U_0. This one-parameter group containing x is unique (3.5.1). It will be denoted by $x(t)$. Furthermore, this notation will mean that 1) in the range of definition t is a group-parameter: $x(t)x(t') = x(t + t')$, and 2) that $x(1) = x$. With this normalization the group-parameter is unique; without it the parameter is determined to within a constant multiple (by the theorem that a continuous, real, linear function of the real variable t is a constant multiple of t).

Let $x(t)$, $y(t) \subset U_0$, $|t| \leq 1$. Then $x(1/m)$, $y(1/m) \varepsilon A_m$ for every $m \varepsilon I$ and for $i \leq m$, (3.7.2),

$$(x(1/m)y(1/m))^i \varepsilon A_m^{2m} \subset U_0^{2k_0}$$

and there is a subsequence $I' \subset I$ such that

$$\lim (x(1/m)y(1/m))^m = z$$

exists as m increase without limit, $m \varepsilon I'$.

The reader will recognize the analogy of the operation $\lim (x(1/m)y(1/m))^m$ to the formula for $\exp (A + B)$ for the exponential map (2.16) taking M_n into Gl_n. In this section we shall work out some of the preliminary steps to showing that there is a vector space which is mapped into G by a completely analogous map. As we shall see in the next section the desired vector space is the space of all homomorphic maps of the reals into G.

3.9.1.   Given local groups $x(t)$, $y(t)$ in U_0 and a sequence $I' \subset I$ such that the following limit exists:

$$z = \lim (x(1/m)y(1/m))^m$$

we shall soon need to know that z belongs to a one-parameter group; ultimately it will be clear that $\lim (x(t/m)y(t/m))^m$ exists and forms such a group. But for the present it is enough that for some subsequence $I'' \subset I'$

$$\lim (x(1/m)y(1/m))^{[mt]} = z(t), \quad |t| \leq 1$$

exists for every t. Although on the face of it, this appears to be an altogether different group from the first one indicated, it will turn out that the two groups coincide.

The desired subsequence I'' may be found by the requirement that $z(t)$ is defined for all rational t. It is easy to see that if $x(1/m)$, $y(1/m)\ \varepsilon\ A_{mk_0}$, then all elements $(x(1/m)y(1/m))^i\ \varepsilon\ U_0^2$ for $i \leq m$. Then we know by Lemma 2 of 3.6 that the set $z(t)$ exists and is a group uniquely determined by $z(1)$.

A convenient way of making sure that $x(1/m)$, $y(1/m)\ \varepsilon\ A_{mk_0}$ is to require that $x(t)$, $y(t) \subset U_0$ for $|t| \leq k_0$. This will also suffice to make available the work of 3.8.2.

3.9.2. Corresponding to the opening remarks in 2.16.3 about abelian subgroups of Gl_n we have the following Lemma.

LEMMA. *If $x(t)$, $y(t)$ are one parameter local groups such that the elements of one commute with elements of the other then*
$$(x(1/m)y(1/m))^m = x(1)y(1) = \lim (x(1/m)y(1/m))^m.$$
The proof is obvious.

3.9.3. The next Lemma is concerned with the commutative character of the operation $\lim (x(1/m)y(1/m))^m$.

LEMMA. *If $x(t)$, $y(t)$ are given local groups and if for $m\ \varepsilon\ I' \subset I$, $\lim (x(1/m)y(1/m))^m$ exists then so does $\lim (y(1/m)x(1/m))^m$ and the two limits are equal.*

This follows from the fact that for any sequence of elements a_m, b_m converging to e for which $\lim (a_m b_m)^m$ exists,
$$\lim (a_m b_m)^m = \lim a_m (b_m a_m)^m a_m^{-1} = \lim (b_m a_m)^m.$$

3.9.4. In the next few Lemmas we are concerned with the question of continuity in the variables x, y appearing in $\lim (x(1/m)y(1/m))^m$.

LEMMA. *Let $x_n(t)$, $n\ \varepsilon\ I$, $x(t)$ and $y(t)$, be local groups, $|t| \leq k_0$. Suppose that*

1) $\lim x_n = x$

and that for all m of a sequence I' and all $n\ \varepsilon\ I$

2) $\lim_{m\ \varepsilon\ I'} (y(1/m)x_n(1/m))^m = z_n$

3) $\lim\limits_{m \,\epsilon\, I'} (y(1/m)x(1/m))^m = z_0.$

Then

$$\lim z_n = z_0.$$

If the Lemma is not true then there is a sequence $I'' \subset I$ such that

$$\lim\limits_{n\,\epsilon\,I''} z_n = z^* \neq z_0.$$

Now for each $n \,\epsilon\, I''$ we may choose an integer $m_n \,\epsilon\, I'$ such that for the sequence of integers m_n

$$\lim\limits_{n\,\epsilon\,I''} (y(1/m_n)x_n(1/m_n))^{m_n} = z^*;$$

this is possible because of 2). Now, because of 3), since $m_n \,\epsilon\, I'$,

$$\lim\limits_{n\,\epsilon\,I''} (y(1/m_n)x(1/m_n))^{m_n} = z_0.$$

Set $y(1/m_n) = a_n,\ x_n(1/m_n) = b_n,\ x(1/m_n) = c_n.$ Then

$$a_n^{m_n} = y(1),\quad b_n^{m_n} = x_n(1),\quad c_n^{m_n} = x(1),$$

and, because of 1),

$$\lim b_n^{m_n} = x(1).$$

Remember that $x(1/m),\ y(1/m) \,\epsilon\, A_{mk_0}.$ Then it follows from 3.8.2 that

$$\lim (a_n b_n)^{m_n} = \lim (a_n c_n)^{m_n}$$

and this means that

$$z^* = z_0.$$

The contradiction completes the proof.

COROLLARY. *Suppose that $x_n(t)$, $y(t)$ are as above, with $\lim x_n = x$ and suppose that z_n is defined for all n as in 2). Then $\lim z_n$ exists, and 3) is satisfied by setting $z_0 = \lim z_n$.*

This follows from the Lemma by the type of argument of 3.6.1.

3.9.5. Next we consider continuity in both variables.

LEMMA. *Let $x_n(t)$, $y_n(t)$, $n \,\epsilon\, I$, and $x(t)$, $y(t)$, $|t| \leq k_0$, be local* *groups in U_0 with* $\lim x_n = x$ *and* $\lim y_n = y$. *Suppose there is a sequence* $I' \subset I$ *such that for all* $n \varepsilon I$

1) $\lim_{m \varepsilon I'} \left(y_n(1/m)x_n(1/m) \right)^m = z_n$

and

2) $\lim_{m \varepsilon I'} \left(y(1/m)x(1/m) \right)^m = z_0.$

Then

3) $\lim z_n = z_0.$

For each n consider the sequence for $m \varepsilon I'$

$$\left(y(1/m)x_n(1/m) \right)^m.$$

Choose a sequence $I'' \subset I'$ such that

$$\lim_{m \varepsilon I''} \left(y(1/m)x_n(1/m) \right)^m = z_n^*$$

exists for each n.   It follows from the preceding Lemma that

$$\lim_n z_n^* = z_0.$$

For some sequence of integers $m_n \varepsilon I^* \subset I''$

$$\lim_{n \to \infty} \left(y(1/m_n)x_n(1/m_n) \right)^{m_n} = z_0.$$

For this sequence let $m' = m_n$ and let

$$a_{m'} = y_n(1/m_n), \ \ b_{m'} = x_n(1/m_n), \ \ c_{m'} = x(1/m_n).$$

Then $\lim (b_{m'})^{m'} = x = (c_{m'})^{m'}$ and consequently, for this subsequence

$$\lim z_{m'} = z_0.$$

Since z_0 is independent of I^*, we may be sure that $\lim z_{m'} = z_0$ when m' ranges over all of I''. Since z_0 is also independent of I'', even though the elements z_n^* depend on I'' we may be sure (3.6.1) that $\lim z_n = z_0$. This completes the proof.

COROLLARY. *Let us be given local groups* $x_n(t)$, $y_n(t)$ *as above with* $\lim x_n = x$ *and* $\lim y_n = y$. *Suppose that* z_n *is defined as in* 1). *Then* $\lim z_n$ *exists and* 2) *is satisfied by setting* $z_0 = \lim z_n$.

This follows from the Lemma and the standard device (3.6.1).

3.9.6. We shall now show that when $\lim (x(t/m)y(t/m))^m$ is defined on a dense set of values of t in some range then it can be extended to all t in the range and we shall show that it is continuous in t.

LEMMA. *Let $x(t)$, $y(t)$ be one-parameter groups in U_0 for $|t| \leq k_0$, let $t_n \, \varepsilon \, [0, 1]$ and let $\lim t_n = t$. Suppose that we are given a sequence I' such that*

$$1) \quad \lim_{m \, \in \, I'} (x(t_n/m)y(t_n/m))^m = z_n, \quad n \, \varepsilon \, I.$$

Then, for m over the same sequence I',

$$\lim (x(t/m)y(t/m))^m = z$$

exists and

$$\lim z_n = z.$$

For the proof, set $x_n(s) = x(t_n s)$ and set $y_n(s) = y(t_n s)$, for each n; and apply 3.6.1 and the last Corollary. This concludes the proof of the Lemma. Then if 1) happens to be defined on a dense subset of a domain, it can be extended to the whole domain by continuity.

3.9.7. The next Lemma is concerned with the property of associativity.

LEMMA. *Let $x(t)$, $y(t)$, $z(t)$ belong to U_0 for $|t| \leq k_0$ and suppose that for a sequence $m \, \varepsilon \, I'$*

$$\lim (y(1/m)z(1/m))^m = v$$

and

$$\lim (x(1/m)v(1/m))^m = w$$

exist. Then for the same sequence

$$\lim (x(1/m)y(1/m)z(1/m))^m = w.$$

The assumption which is made in the statement of the Lemma that v is on a one-parameter group so that $v(1/m)$ is defined for all m is justified by 3.9.1. Let $a_m = x(1/m)$, let $b_m = y(1/m)z(1/m)$, and let $c_m = v(1/m)$. We are given that $\lim b_m^m = \lim c_m^m = v$. One can see that $v(1/2m) \,\epsilon\, A_m$, and then $v(1/m) \,\epsilon\, A_m^2$, and 3.8.2 applies. Let $m^* = [m/k_0]$, as in 3.8.

Now for every $i \leq m$, $(x(1/m)y(1/m)z(1/m))^i \,\epsilon\, U_0^3$ and we shall suppose (ultimately using 3.6.1) that $(a_m b_m)^m$ converges and so do the sequences $(a_m b_m)^{m^*}$ and $(a_m c_m)^{m^*}$ (compare 3.9.1), and therefore by 3.8.2

$$\lim \, (a_m b_m)^{m^*} = \lim \, (a_m c_m)^{m^*}.$$

This common limit is $w(1/k_0)$. If we raise the terms of this equation to the k_0-th power we obtain the desired result, proving the Lemma.

3.9.8.   If a positive integer n is given we can find a neighborhood of e where the associative principle of 3.9.7 can be extended to any n one-parameter groups contained in that neighborhood. Thus, for each $n \,\epsilon\, I$, let L_n be defined as follows:

$$L_n = \{x; \; x \,\epsilon\, x(t) \subset U_0 \text{ for } |t| \leq (2k_0)^n\}.$$

If $x \,\epsilon\, L_n$ then $x(1/m) \,\epsilon\, A_{m''}$, where $m'' = m(2k_0)^n$. If $x, y \,\epsilon\, L_n$ and if $z = \lim \big(x(1/m)y(1/m)\big)^m$, $m \,\epsilon\, I^* \subset I$, then by 3.9.1, z is on a one-parameter group in U_0; by the argument there using the definition of k_0 (3.7.4) it follows that $z \,\epsilon\, L_{n-1}$. Now if $x_1, \ldots, x_n$ are elements of L_n we can apply 3.9.7 to x_1, x_2, x_3 to get an element of L_{n-2}, and by induction we can "associate" all of the x's to a single element of $L_0 \subset U_0$.

If $x(t)$ is any one-parameter local group in G then there is a sufficiently large integer N so that $x(t') \,\epsilon\, L_n$ if $|t'| < 1/N$, and in this sense, at least, L_n is a neighborhood of e with respect to one-parameter local groups in G. That is all we need to know now about L_n; later it will turn out that L_n is really a neighborhood of e in the standard sense.

3.9.9.   For the moment let F denote a countable dense subset of L_1 (3.9.8) and let $I^* \subset I$ denote a sequence such that $\lim \big(x(1/m)y(1/m)\big)^m$ exists for every $x, y \,\epsilon\, F$. Then if $x, y \,\epsilon\, L_1$

are given there exist sequences x_n, $y_n \varepsilon F$ such that $\lim x_n = x$ and $\lim y_n = y$. By 3.9.5

1) $\lim (x(1/m)y(1/m))^m$ exists for m ranging over I^*.

For $n \varepsilon I$, x, $y \varepsilon L_n$ and $r \varepsilon [-1, 1]$ define $z(r) \varepsilon L_{n-1}$:

$$z(r) = x(r) \circ y(r) = \lim (x(r/m)y(r/m))^m, \qquad m \varepsilon I^* \text{ above.}$$

3.9.10. Suppose that x, $y \varepsilon L_2$ and that s, t, $s + t \varepsilon [-1, 1]$. Then using the notation above, and applying 3.9.2, 3.9.7, 3.9.3, etc., we get:

$$\begin{aligned}
z(s + t) &= x(s + t) \circ y(s + t) \\
&= (x(s) \circ x(t)) \circ (y(s) \circ y(t)) \\
&= x(s) \circ (x(t) \circ y(s)) \circ y(t) \\
&= x(s) \circ (y(s) \circ x(t)) \circ y(t) \\
&= (x(s) \circ y(s)) \circ (x(t) \circ y(t)) \\
&= z(s) \circ z(t).
\end{aligned}$$

Now, for $s = t$, the one-parameter groups determined by $z(s)$ and $z(t)$ coincide and using 3.9.2 we get $z(1) = z(1/2) \circ z(1/2) = z(1/2)^2$. By induction, $z(1/2^m)$ is the 2^m-th root of $z(1)$ and hence all these elements lie on the same local group. Similarly, this group contains all elements $z(r)$, for $r = n/2^m \leq 1$. Finally by 3.9.6 it contains all $z(r)$, $r \varepsilon [0, 1]$, and by 3.9.2 $z(s) \circ z(t) = z(s)z(t)$ showing that $z(r)$ is a one-parameter local group.

3.10. The exponential map

Let V_1 denote the group of the reals and let M be the collection of homomorphic maps of V_1 into G. We shall topologize M and we shall define an operation of addition on its elements. Then we shall show that M becomes a finite-dimensional *vector-space*.

Then we can easily define a map, which because of 3.9 it is reasonable to call the *exponential map* (2.15), of M into G.

The elements $f, \ldots$, of M are the continuous homomorphisms $f(t) \,\epsilon\, G$ defined for all $t \,\epsilon\, V_1$:

1) $f(t)f(t') = f(t + t'), \quad t,\ t' \,\epsilon\, V_1.$

These maps are not necessarily isomorphisms, in fact one element of M, denoted by f_0, is defined by

2) $f_0(t) = e$, for all t.

Observe that if $x(t)$, $|\,t\,| \leqq c$ for some $c > 0$, is a one-parameter local group in G, then $x(t)$ has a unique extension to an element of M. It is easy to get an extension (compare 2.16.1); the requirement that the extension is to be a continuous homomorphism makes it unique. It is clear that if $x(t)$ and $x'(t)$ are local groups which coincide for some domain $|\,t\,| \leqq c$, $c > 0$, then they give rise in this way to the same element of M.

The fact that G has non-trivial one-parameter groups shows that M is not a trivial space. For example, if $x(t) \,\epsilon\, U_0$ for $t \,\epsilon\, [-1, 1]$ and $x(t)$ is not identically equal to e, then there exist distinct elements $f_r \,\epsilon\, M$ for all real r defined by

3) $f_r(1) = x(r),$

where 1 denotes the real number 1 of V_1 (which is regarded as fixed when V_1 is given). It may as well be observed at this moment that 3) also defines a map of M into G (see 17) below).

To topologize M we define a metric $d(f, g)$, $f,\ g \,\epsilon\, M$, using any convenient metric $d'(x, y)$ in G. Thus

4) $d(f, g) = \text{l.u.b. } \{d'(f(t),\ g(t)),\ |\,t\,| \leqq 1\}$

Because f and g are continuous maps into G, $d'(f(t),\ g(t))$ is a continuous function of t and has a maximum on a compact range. Then $d(f, g)$ is defined. For each $f,\ g \,\epsilon\, M$ and for some $t' \,\epsilon\, [-1, 1]$ depending on f and g

4') $d(f, g) = d'(f(t'),\ g(t')).$

Then given $f,\ g,\ h \,\epsilon\, M$, and t' satisfying 4'), we see that

$$d(f, h) + d(h, g) \geqq d'(f(t'), h(t')) + d'(h(t'), g(t'))$$
$$\geqq d'(f(t'), g(t')) = d(f, g).$$

The symmetry of $d(f, g)$ is obvious and it is also obvious that $d(f, f) = 0$. If $d(f, g) = 0$ for some $f, g \varepsilon M$ then $f(t)$ and $g(t)$ coincide for all $t \varepsilon [- 1, 1]$, and it follows from 1) that $f(t)$ and $g(t)$ coincide for all t, and so $f = g$. This concludes the demonstration that $d(f, g)$ is a metric; and M is now a metric space.

Since U_0 is a neighborhood of e in G there is also a metric neighborhood of e contained in U_0. Choose a number $d^* > 0$ such that if $d'(x, e) < d^*$, then $x \varepsilon U_0$.

We show next that the set of points

5) $\quad M_0 = \{f; d(f_0, f) \leqq d^*\} \subset M$

is compact. It follows from the choice of d^* that if $f \varepsilon M_0$ then $f(t) \subset U_0$ for $|t| \leqq 1$. Now if $f_n \varepsilon M_0$, $n \varepsilon I$, is given we can find a subsequence I' such that the elements $f_n(1)$ converge to an element x and then the one-parameter groups $f_n(t)$, $n \varepsilon I'$, converge to a one-parameter local group $x(t)$, $|t| \leqq 1$. This group has an extension to an element f of M and the sequence f_n converges to f in the sense of the metric on M. Then every sequence in M_0 has a convergent subsequence and it follows (1.10.6) that M_0 is compact.

Next we define the addition of pairs of elements of M. Given $f, g \varepsilon M$ there is a positive integer N such that $f(t), g(t) \varepsilon L_4$ for $|t| < 1/N$. Now, following 3.9.10, we define a one-parameter local group $z(t)$ by

$$z(t) = f(t) \circ g(t), |t| < 1/N.$$

Let us denote by $h(t)$ the unique extension of this local group to an element of M. It is easy to see that only the domain of definition of $z(t)$ is affected by the integer N, and therefore $h(t)$ is independent of N. This defines the addition:

6) $\quad f + g = h.$

It now follows from 3.9.3 that for every $f, g \varepsilon M$

7) $\quad f + g = g + f.$

It is clear that the element f_0 is the zero of this addition:

8) $f_0 + f = f = f + f_0.$

Since $f(t)f(-t) = e$, we denote by $-f$ the element $f(-t) \varepsilon M$, and then

9) $f + (-f) = f_0 = (-f) + f.$

If for any f, $g \varepsilon M$, $f + g = f_0$, then $f(t)g(t) = e$ for all t and $g = -f$; this follows from the fact that $\lim (f(t/m)g(t/m))^m = e$ implies (by 3.8.1, t small enough) that $\lim (f(t/m))^m (g(t/m))^m = e$, and this gives $f(t)g(t) = e$. Therefore each f has a unique inverse; equivalently, 9) has a unique solution.

It is easy to see that

10) $(f + g) + h = f + (g + h);$

because it follows from 3.9.7 that $f(t) \circ g(t) \circ h(t)$ is associative for small enough t. Finally, M is a vector space, using the reals as coefficients. Thus for every real r and every $f \varepsilon M$ there is defined rf:

11) $rf = rf(t) = f(rt), \quad t \varepsilon V_1,$

and, we obtain $r(f + g) = rf + rg$ and,

12) $(r+s)f = f((r+s)t) = f(rt)f(st) = f(rt) \circ f(st) = rf+sf.$

Concerning the continuity of the algebraic operations in M we observe first that if f, $f_n \varepsilon M$ then $\lim f_n = f$ means that we have $\lim f_n(t) = f(t)$ uniformly for $|t| \leq 1$. Therefore, if $\lim f_n = f$ there exists a positive integer n' such that $f(t)$, $f_n(t) \varepsilon U_0$ for $|t| \leq 1/n'$, for all n. If we are also given real numbers r, r_n with $\lim r_n = r$ then $f_n(r_n t)$ and $f(rt)$ belong to U_0 for sufficiently large n and for all t numerically less than $1/2n'$ if $|r| \leq 1$ and all t numerically less than $|1/2rn'|$ if $|r| \geq 1$. Then it follows from 3.6.7 that $f_n(r_n t)$ converges to $f(rt)$ for all t in the indicated range. It follows that this convergence holds uniformly for $|t| \leq 1$ and therefore

13) $\lim f_n = f$, $\lim r_n = r$ implies $\lim r_n f_n = rf.$

If f, f_n, g, $g_n \, \varepsilon \, M$ are given and $\lim f_n = f$, $\lim g_n = g$ then for some positive integer n', $f(t)$, $f_n(t)$, $g(t)$, $g_n(t) \, \varepsilon \, L_2$ (3.9.8) provided $|\,t\,| \leq 1/n'$. Now using the notation of 3.9.10, it follows from 3.9.5 that $f_n(t) \circ g_n(t)$ converges to $f(t) \circ g(t)$. This convergence holds uniformly for $|\,t\,| \leq 1$, so that

14) $\lim f_n = f$, $\lim g_n = g$ implies
       $\lim (f_n + g_n) = f + g$.

This concludes the demonstration that M is a locally compact vector space. It follows from the theorem of Riesz that M must have a finite basis

15) $f_1, \ldots, f_N$

and every $f \, \varepsilon \, M$ is uniquely representable as

16) $f = r_1 f_1 + \ldots + r_N f_N$

with real coefficients $r_1, \ldots, r_N$. We do not need to prove this directly since we can see it quite easily from the theorem of Pontrjagin (2.21). We have proved that M is a locally compact abelian group and that each element lies on a one-parameter group, by 12) above. Then M is connected (since these groups contain f_0). If $f \, \varepsilon \, M$, the collection $\{nf\}$, $n \, \varepsilon \, I$, has no limit element in M and consequently M has no compact subgroup, and therefore no "small subgroups", except e. Now it follows from 2.21 that M is isomorphic to a vector space V_N for some $N \geq 1$.

There is a map of M into G defined by

17) $f \to f(1)$

and it follows from the way in which M was topologized that this map is continuous ($\lim f_n = f$ implies $\lim f_n(1) = f(1)$). If $x \, \varepsilon \, L_0$ (3.9.8) then as we remarked above there is one and only one $f \, \varepsilon \, M$ such that $f(1) = x$, and $f(t) \, \epsilon \, U_0$, $|\,t\,| \leq 1$. Thus, 17) is a one-one map of a neighborhood of the zero of M onto L_0; and similarly for L_N, N as in 15).

It will be seen later that L_0 (and also each L_n, $n \, \varepsilon \, I$) is a neighborhood of e in G (supposing that G is connected). The example of the map of M_n into Gl_n (2.16) would lead one to expect that any (minimal) basis of vectors in M (for example, the set in 15) would give rise to an analytic coordinate scheme in G in a neighborhood of e. This turns out to be true.

3.11. The adjoint representation

We now show that G is a topological transformation group of M in a way which is completely analogous to the adjoint representation of a Lie group (compare the discussion of Gl_n in 3.0). Although the vector space M (3.10) is naturally associated with the identity-component of the group G (with no small subgroups), it is immaterial in defining the present representation of G whether or not G is connected.

Let a be an arbitrary momentarily fixed element of G. Let f denote an arbitrary element of M. Then $a^{-1}f(t)a \, \varepsilon \, L_0$ for all sufficiently small t, and $a^{-1}f(t)a$ has a unique extension to an element f_a of M. This gives a transformation of M into itself associated with the element a:

18) $a \, \varepsilon \, G : f \to f_a \, \varepsilon \, M.$

Of course, we have it in mind that for arbitrary real r and s

18') $f_a(r+s) = a^{-1}f(r+s)a = a^{-1}f(r)aa^{-1}f(s)a = f_a(r)f_a(s)$

so that f_a is indeed an element of M. Now if r is any real number:

19') $a^{-1}f(rt)a = f_a(rt)$

for all t; this follows from 18' for integral r, and then for rational r and finally it is true for all r by continuity. Writing 19') in accordance with 11) we have:

19) $(rf)_a = rf_a.$

Next, for given $f, g \, \varepsilon \, M$ and for all $t \, \varepsilon \, V_1$ and all $m \, \varepsilon \, I$

20') $a^{-1}(f(t/m)g(t/m))^m a = (a^{-1}f(t/m)aa^{-1}g(t/m)a)^m$
$= (f_a(t/m)g_a(t/m))^m.$

For sufficiently small t, 20') gives

20'') $a^{-1}(f(t) \circ g(t))a = f_a(t) \circ g_a(t).$

This holds for all t and we get

20) $(f + g)_a = f_a + g_a$

Given $a, b \varepsilon G$, it is clear that for all $f \varepsilon M$ and all t,

21') $b^{-1}a^{-1}f(t)ab = (ab)^{-1}f(t)ab;$

this gives

21) $f_{ab} = (f_a)_b,$

and shows that G is a transformation group of M. It follows from 19) and 20) that 18) is an endomorphism of M. Then 21) shows that 18) defines an automorphism (that is, the map is not merely linear and into but also one-one onto as one sees by taking $ab = e$ in 21)).

Each map: $f \to f_a$ is continuous; this follows from the fact that if $f, f_n \varepsilon M$ and $\lim f_n = f$, then $\lim a^{-1}f_n(1)a = a^{-1}f(1)a$. But if moreover $a, a_n \varepsilon G$ and $\lim a_n = a$ then

21'') $\lim a_n^{-1} f_n(1)a_n = a^{-1}f(1)a.$

It follows that G is mapped continuously as well as homomorphically into the group Gl_N.

Let $H \subset G$ denote the kernel of the map 18); that is, $a \varepsilon H$ if and only if f_a is always equal to f. An equivalent, useful criterion is the following:

22) $a \varepsilon H$ if and only if $a^{-1}f(t)a = f(t)$ for all $f \varepsilon M$ and all t.

The kernel H is a closed invariant subgroup of G (1.13.1) and G/H is mapped in a one-one way upon some subgroup of Gl_N. The natural map of G onto G/H is open, the map of G into Gl_N is continuous. It follows that the map of G/H into Gl_N is continuous as well as one-one.

Now it follows from 2.16 that G/H is isomorphic to a Lie group. If it happens to be the case that G has no invariant subgroups, then G will be isomorphic to a Lie group. At the other extreme, if G happens to be abelian then G is a Lie group (2.21).

3.12. The kernel of the representation

We shall prove now that H (3.11) is a Lie group whose identity component H_0 is abelian, and in a later section we shall see that H is a central subgroup of G, if G is connected. We need the following Lemma.

LEMMA. *Let F be a topological group and let K be a central closed subgroup such that F/K is a one-parameter group. Then F is an abelian group.*

Let a_0 be an arbitrary element of F. The coset a_0K has a 2^n-th root, a_nK, $n = 1, 2, \ldots$:

$$(a_nK)^{2^n} = a_0K.$$

Therefore, for each n we can find an element $k \,\varepsilon\, K$, depending on n, such that:

$$ka_n^{2^n} = a_0.$$

This shows that a_0 commutes with every a_n and therefore also with every element of every a_nK. Therefore a_0 commutes with the group generated by the sets a_nK, and it is easy to see that this group is dense in F. Since a_0 is an arbitrary element of F, this concludes the proof.

3.12.1.   We return to the group H. Suppose first that H is totally disconnected. In this case, since H contains no small subgroups, H must be discrete (2.3). Then H is a Lie group by definition and, since G and G/H are locally isomorphic, G is a Lie group.

Next we suppose that the identity component H_0 of H is not trivial. Then the closed group K generated by one-parameter subgroups of H_0 is not trivial and is in the center of H, by the nature of H. If K did not coincide with H_0, then H_0/K would contain a one-parameter subgroup (2.22). In this case H_0 would contain a subgroup F such that F/K was a one-parameter group. By the preceding Lemma, F is abelian, and its closure is abelian (1.11.3). Since $\overline{F}$ is in H_0, it has no small subgroups, and so F is an abelian Lie group every element of which lies on a one-parameter subgroup (2.16.1). It follows that F is in K, by definition of K, and this contradiction shows that K must coincide with H_0; then H_0 is an abelian Lie group.

Finally we must show that H/H_0 is discrete. Let H' be an open subgroup of H such that H'/H_0 is compact. Let D be a discrete subgroup of H_0 such that H_0/D is compact (2.21). It follows easily from

$$H'/H_0 = (H'/D)/(H_0/D)$$

that H'/D is compact. Since H'/D is locally isomorphic to H' (and H' does not have small subgroups except e) it follows that H'/D is a Lie group, and its identity-component must be an open subset. Since H and H'/D are locally isomorphic we see now that H_0 must be open and that H/H_0 is discrete. This completes the proof that H is a Lie group.

Since

$$G/H = (G/H_0)/(H/H_0),$$

it follows that G/H_0 is *locally isomorphic* to a matrix group. In the next chapter we shall see that H_0 is central and then conclude the demonstration that G is a Lie group.

CHAPTER IV

Approximation by Lie Groups

4.0. Introduction

A group G is said to be approximated by Lie groups if every neighborhood of the identity contains an invariant subgroup H such that G/H is isomorphic to a Lie group. In this chapter we shall prove that every locally compact group G has an open (and closed) subgroup G' such that G' can be approximated by Lie groups. The group G' is not uniquely determined but is merely any open subgroup of G such that G'/G_0 is compact, G_0 being the identity component. An occasionally useful property (see 2.5) of G' is that it is a subgroup of G generated by a compact neighborhood (symmetric) of the identity.

In Chapter II we showed that every *compact* group can be approximated by Lie groups. There the approximating groups G/H were shown to be Lie groups from the fact that they were isomorphic to matrix groups. A non-compact Lie group is not necessarily isomorphic to a matrix group and the fact that it is locally isomorphic to such a group is difficult to prove, in general. We do not prove this fact here. We have shown in the preceding chapter that a locally compact group without small subgroups has an invariant Lie group H with an abelian identity component such that G/H is isomorphic to a matrix group. We shall show next that such a group is a Lie group. Then we shall show that every locally compact G such that G/G_0 is compact has arbitrarily small invariant subgroups H such that G/H has no small subgroups, — therefore is a Lie group.

4.1. One-parameter factor groups

If G is a topological group and H is a closed invariant subgroup such that H and G/H are Lie groups, then G is a Lie group. This was first proved by Chevalley [2] for the case, essentially, that H and G/H are abelian. In the form in which we shall use it for our exposition it was proved by Kuranishi [1] (the general case was obtained by Iwasawa [1] and also by Gleason [4]). Although we do not give a direct proof of the general result, this will of course follow from the theorems on the structure of locally compact groups given below. The proof of the Kuranishi result requires that we find in G a set which locally cross-sections the cosets xH. Such a set is locally homeomorphic to G/H and permits us to carry coordinates from G/H to coordinates in the local cross-section. These coordinates, supplemented by the coordinates of H will be used to construct coordinates for G.

It will be convenient to have a Lemma proved by Iwasawa [1], which is Lemma 2 below. We begin with a fact about toral groups.

Lemma 1. *Let G be a connected topological group containing a closed invariant subgroup H which is a finite-dimensional toral group. Then H is central.*

Since G is connected it is generated by any neighborhood of the identity (1.24). Therefore to prove that H is central it will be sufficient to prove that there is a $U(e)$ in G such that every element in U commutes with every element in H.

If g is any element in G then g determines an automorphism of H by the following:

$$g : h \to ghg^{-1},$$

and because H is compact this automorphism can be made arbitrarily small provided U is chosen properly and g is in U.

The proof will be completed by showing that any small automorphism of H is the identity. The reader can verify this when H is the circle group. Assume now that H is not the circle group, that is dim $H = n > 1$.

We may write

$$H = C_1 \times T_{n-1},$$

where C_1 is a circle group and T_{n-1} is a toral group of dimension one less than H. By an automorphism of H, C_1 is carried to a group C_1' in H,

$$C_1' \subset C_1 \times T_{n-1},$$

and we may project C_1' into T_{n-1} and there obtain a group $C_1'' \subset T_{n-1}$.

If the automorphism is small, the group C_1'' will be near the identity of T_{n-1}, but the only group in T_{n-1} near the identity is the group $\{e\}$. Hence any small automorphism of H leaves C_1 invariant. The same is true for each C_i, where

$$H = C_1 \times \ldots \times C_n.$$

Hence every sufficiently small automorphism of H is the identity and this completes the proof of the Lemma.

4.1.1. REMARK: Under the same hypothesis except that $H = T \times V$, where T is a finite dimensional torus and V a vector group it follows that T is central. The proof above can be extended without change to this case; *or* one can observe that if H is invariant then T must be invariant too.

4.1.2. As the term is used below a one-parameter group is a topological group which is the continuous homomorphic image of the reals; hence it may or may not be locally compact and both situations will be needed in the applications. Thus also the group K of the next Lemma may or may not be locally compact.

LEMMA 2. *Let K be a topological group and H an invariant closed subgroup. Suppose further that H is a locally compact connected abelian group without small subgroups and that K/H is a one-parameter group. Then K contains a one-parameter group X such that $K = XH$ and $X \cap H = e$.*

Although K need not be locally compact note that the inverse of any arc in K/H is locally compact (similar to 2.2); hence K is the union of a countable increasing family of compact subsets. Recall that H is the product of a torus and a vector space.

Following Iwasawa we proceed inductively on the dimension

of H (note that the Lemma is true when dim $H = 0$), and we assume the Lemma to be true for subgroups of dimension less than dim H.

Suppose first that H contains a connected closed subgroup H_1 which is invariant in K, $0 < \dim H_1 < \dim H$. Then since (up to isomorphism)

$$K/H = (K/H_1)/(H/H_1),$$

and H/H_1 is of lower dimension than H, it follows by assumption that K/H_1 contains a one-parameter group V of the type required in the conclusion of the Lemma.

This gives a group K' in K such that $K'/H_1 = V$. Since H_1 is of lower dimension than H we may assume that K' has a one-parameter group X satisfying $K' = XH_1$, $X \cap H_1 = e$. But because $K' \cap H = H_1$ this group X also satisfies

$$K = XH, \ X \cap H = e.$$

Therefore the general case of the Lemma has been reduced to the case where no proper connected subgroup of H is invariant in K.

By the remark in 4.1.1, H is therefore either a torus *or* a vector group. In case H is a torus, we know that H is central by Lemma 1 above, and the central case will be considered later. The case considered next is the one where H is a vector group, which is not central.

We observe that any g in K determines an automorphism of the group H given by $g : h \to g^{-1}hg$. This is an automorphism of H in the vector space sense as well as in the group sense and is a linear transformation of H.

Let x_0 be an element of K which does not commute with all elements of H. Let X denote the subgroup of elements of K which commute with x_0.

Acting on elements u of H, x_0 gives a linear transformation

$$x_0^{-1}ux_0u^{-1} = v \ \epsilon \ H$$

of the n-dimensional vector space H into itself. This transformation is equivalent to a system of n linear equations with constant coefficients in the coordinates of H. If $v \in H$ is given, the equations can be solved for $u \in H$ provided that the determinant of the system of equations is not zero — or equivalently provided that the homogeneous equations

1) $x_0^{-1} u x_0 u^{-1} = e$

cannot be solved except by $u = e$.

We now investigate the equations 1). Let H' denote the space of solutions of 1), that is $u \in H'$ if and only if $x_0 u = u x_0$. The set H' is a linear subspace of H.

Let x be an element of K. Since K/H is abelian,

$$x_0 x = w x x_0, \quad w \in H.$$

Then for $u \in H'$, remembering that $w^{-1} u w = u$,

$$x^{-1} u x = x^{-1} x_0^{-1} u x_0 x$$
$$= (x_0 x)^{-1} u (x_0 x)$$
$$= (w x x_0)^{-1} u (w x x_0)$$
$$= x_0^{-1} x^{-1} u x x_0.$$

Therefore $x^{-1} u x$ belongs to H'. It follows now that H' is invariant in K. Therefore $H' = H$ or $H' = e$. But $H' \neq H$ since some element in H does not commute with x_0. This concludes the proof that given $v \in H$

2) $x_0^{-1} u x_0 u = v$

can always be solved for u.

Finally let y belong to K. We shall show that x_0 commutes with one element on the coset yH. Define $v \in H$ as follows:

$$v = x_0^{-1} y x_0 y^{-1}$$

and solve

$$v = x_0^{-1} u x_0 u^{-1}$$

for $u \in H$. We have

$$y x_0 y^{-1} = u x_0 u^{-1}$$

so that

$$u^{-1}yx_0 = x_0 u^{-1}y$$

and x_0 commutes with $u^{-1}y$ which belongs to $Hy = yH$. If x_0 could commute with two distinct elements in yH it would commute with an element of H distinct from e. Therefore there is just one element on each coset yH which commutes with x_0. The set X of these elements is the desired one-parameter group isomorphic to K/H, as we shall now see.

We know there is a continuous one-one homomorphism T (the natural map restricted to X) taking X onto K/H. The group K is the union of a countable increasing family of compact sets A_n and hence K/H is the union of the increasing compact sets $T(X \cap A_n)$. Further T is a homeomorphism on each of the sets $X \cap A_n$. By a category argument some interval in K/H is contained in some set $T(X \cap A_n)$, n fixed. Hence $X \cap A_n$ contains an interval mapped homeomorphically onto an interval in K/H. From this it can be deduced that X is a one-parameter group as desired (compare the argument two paragraphs down).

In the application of this Lemma which we are about to make, K is a subgroup of a group G in which H commutes with all elements on one-parameter subgroups of G (as in 3.11). In that case it follows that H commutes with the elements of X and K is necessarily abelian.

If K is abelian and $x(t)$, $h(t)$ one-parameter groups in K, then the set of elements $z(t) = x(t) \cdot h(t)$ is a one-parameter group. In this way it is clear that every element of K (K as above) is on a one-parameter group. Of course, X or any other one-parameter group in K not in H constitutes a *cross-section* of cosets xH in K.

To complete the proof of the Lemma we need to consider the case that H is a central subgroup of K. In this case it follows from the Lemma in 3.12 that K is abelian. It also follows from the present induction that H is one-dimensional (so that in fact K is a continuous homomorphic image of the vector-plane) but it is not important to take advantage of this. It has to be kept in mind that K is not necessarily locally-compact; in the application which we have to make the group K is generated by a local one-parameter subgroup of a group G/H and K may fail to be a closed subgroup.

Since K/H is a one-parameter group there is a compact interval J around e in K/H which has no proper subgroup (except the identity). Let J^* denote the inverse of $J \subset K$. Now let us retopologize K, if necessary, as follows. If W is a neighborhood of e in K, let $W^* = W \cap J^*$ be a neighborhood in the "new" group K^*. Clearly K is a continuous, one-one image of K^*. Note that J^* is connected and generates K^*. Then K^* is a locally compact connected abelian group, and K^* has no small subgroups and is an abelian Lie group. Therefore K^*, which is the product of a vector group and a toral group (2.16.1), has a one-parameter group X^* such that $K^* = H^*X^*$, and X^* meets H^* in the identity (since $H \subset J^*$, H and H^* are isomorphic). Let $X \subset K$ be the image of X^* in K. Then X is the desired one-parameter group.

4.2. Groups with no small subgroups

We have already proved the larger part of the following principal theorem. We have not proved the existence of the cross-sectioning set S for which the theorem calls, but we prepared the way for that in the preceding Lemma.

THEOREM. *Let G be a locally compact connected group with no small subgroup and let H denote the center of G. Then H is a Lie group locally isomorphic to a vector space of dimension $n \geq 0$ and G/H is a Lie group of dimension $m \geq 0$ isomorphic to a matrix group. Furthermore there is a neighborhood U of e such that each of its points is on a one-parameter group in G. The set consisting of points x on segments in U (a segment is an arc of a one-parameter group joining e and x) contains an m-cell S which is a local cross section of cosets xH.*

The coordinate system obtained in this theorem for S together with coordinates in H will help later in finding analytic coordinates in G.

We begin with the group H constructed in 3.11 and with the

natural map T and the map T^* ($H_0 =$ identity component of H)

$$T^* : G \to G/H_0$$
$$T : G \to G/H.$$

The groups G/H_0 and G/H are locally isomorphic.

At present we know H includes the center of G and we prove next that H is the center of G. Since G/H is mapped by a continuous one-one homomorphism into Gl_N, there is a neighborhood V of the identity of G/H (and a similar V_0 in G/H_0) which is ruled in a unique way by segments of one-parameter groups radiating from e (2.16.2).

Let U_0 be the standard neighborhood of e in G with which we worked in Chapter III (U_0^8 has only the trivial subgroup e), and assume that

$$T(U_0) \subset V, \ T^*(U_0) \subset V_0.$$

We set $U = U_0$ and define L here as the L_m in 3.9.8. The set $H \cap L$ coincides with H locally.

Let x be any element of $U = U_0$. Then $T^*(x)$ is on an arc in V_0 of a one-parameter group say $q(t)$ in G/H_0. Let K be the inverse in G of the one-parameter group $q(t)$. By the previous section K contains a one-parameter group which contains x, and thus every element in U_0 is on a one-parameter group in G.

Because G is a connected group it is generated by U_0. Since every element of U_0 is on a one-parameter group, every element of U_0 commutes with every element of H.

It follows that H is in the center of G. Finally, H having been defined as the group which commutes with all elements on one-parameter subgroups, H coincides with the center of G. This is very convenient in the extension-theorem which is to follow (4.3).

We turn now to the construction of the cross-section S of the cosets xH. We shall get this set quite easily from the connection between the "vector-addition" (3.9) defined in L, and the "vector-addition" defined in G/H (in a neighborhood of the identity, of course) by virtue of its being a matrix group (2.16). If we prefer we can associate a "vector-addition" with G/H because it is a group without small subgroups; but this is merely a longer way to the same "vector-addition" of its elements.

Let $z_1(t)$, $z_2(t)$ be arbitrary one-parameter groups in L and let $Z_1(t)$ and $Z_2(t)$ be their images in G/H:

$$T z_i(t) = Z_i(t), \ i = 1, 2, \ |t| \leq 1.$$

Then we have to show that (notation as in 3.9.9)

$$T(z_1 \circ z_2) = Z_1 \circ Z_2,$$

where vector-addition on the left is $\lim (z_1(1/m)z_2(1/m))^m$, and on the right it is defined by $\lim (Z_1(1/m)Z_2(1/m))^m$ (for some subsequence). But

$$T(z_1(1/m)z_2(1/m))^m = (Z_1(1/m)Z_2(1/m))^m$$

from the fact that T is a homomorphism, and the corresponding limits are equal from the fact that T is continuous.

Finally let:

$$Z_1(t_1), \ldots, Z_m(t_m)$$

be a linearly independent set of one-parameter groups in G/H in the vector-addition generating the vector-space of G/H. Now choose

$$z_1(t_1), \ldots, z_m(t_m)$$

as one-parameter subgroups of G such that $Z_i = Tz_i$, $i = 1, \ldots, m$. Let S contain all elements of the form

$$z = z_1(s_1) \circ z_2(s_2) \circ \ldots \circ z_m(s_m)$$

for sufficiently small real numbers $s_1, \ldots, s_m$. Let z' be another element of S:

$$z' = z_1(s_1') \circ \ldots \circ z_m(s_m').$$

Then for small s_i and s_i'

$$T(z) = T(z')$$

if and only if each s_i equals the corresponding s_i', because the "vectors" $Z_1, \ldots, Z_m$ are independent. It follows that $z = z'$ if and only if the coordinates $s_i = s_i'$, $i = 1, \ldots, m$. This shows that the "vectors" $z_1, \ldots, z_m$ are linearly independent in the vector-addition in L, and it shows that two different elements z and z' of S belong to different local cosets: $zH \cap L \neq z'H \cap L$. On the other hand, since the one-parameter groups $Z_1, \ldots, Z_m$ generate a neigborhood of the identity of G/H, it follows that every coset zH, for z near enough to e, is the image of an element of S. Then we have proved that S is also a local cross-section of H cosets in the neighborhood of e.

This concludes the demonstration of the theorem except to remark that the integer m giving the dimension of G/H is zero if and only if G coincides with H. In this case G is an abelian Lie group and the theorem is true, the cross section S reduces trivially to the identity. The more interesting special case is that H is the identity. Here G is isomorphic to a matrix group and the theorem is true with S trivially equal to G.

COROLLARY 1. *If G is a connected locally compact group without small subgroups there is a neighborhood of e which is uniquely ruled by segments of one-parameter groups.*

We select in $H \cap L$, n independent one-parameter groups

$$x_1(r_1), \ldots, x_n(r_n)$$

generating $H \cap L$. From the fact that S is a local cross-section of H-cosets, there is a neighborhood of e in G such that every element is uniquely represented as a linear combination (in vector-addition) of an element in S and an element of $H \cap L$. This gives to each element of G a unique set of $m + n$ coordinates

$$t_1, \ldots, t_m; \ r_1, \ldots, r_n;$$

and each such element lies on the unique one-parameter local group:

$$t_1 t, \ t_2 t, \ldots, t_m t; \ r_1 t, \ldots, r_n t; \ |t| \leq 1.$$

This proves the corollary.

A system of coordinates of the kind just described is called a *canonical system of the first kind.* It can be shown (though we shall not do so) by use of the results of 4.4 that in some neighborhood of e every element is also uniquely represented as a *product* in the natural multiplication in G:

$$z_1(t_1) \cdot z_2(t_2) \ldots z_m(t_m) \cdot x_1(t_{m+1}) \ldots x_n(t_{m+n}).$$

Such a coordinate system is called *canonical of the second kind.* Of course, it is not meant that an element will necessarily have the same coordinates in these two systems.

There does not seem to be any easy direct proof of the fact that the group multiplication and the inverse-operation are analytic when expressed in canonical coordinates of either first or second kind. In the next section we shall prove that G is a differentiable group of class C^∞.

COROLLARY 2. *If G is any locally compact group without small subgroups (not necessarily connected) then G/G_0 is discrete; hence Corollary 1 is true in this case also.*

For such a general group G it has already been shown that H is a Lie group and that H_0 is abelian (3.12). The groups G/H and G/H_0 are locally isomorphic and therefore G/H_0 is a Lie group and its identity-component is open. This component maps into the identity of G/G_0 under the natural map

$$G/H_0 \to G/G_0.$$

Since this map is open, being part of the chain of maps

$$G \to G/H_0 \to G/G_0,$$

it follows that the identity of G/G_0 is an open set and G/G_0 is discrete. This completes the proof of the Lemma.

We shall show below that G is a Lie group. The converse proposition that a Lie group does not have small subgroups is true, and thus the property of "no small subgroups" is characteristic of Lie groups. The converse will be proved indirectly, — but a direct proof independent of the foregoing can be found in 5.2.

We have shown above that a neighborhood of the identity in a locally compact group without small subgroups is homeomorphic to a euclidean cell. The dimension of this cell is unique (by a theorem of Brouwer) and we can therefore attach a finite dimension to any locally compact group without small subgroups.

4.3. An extension theorem

The first extension theorem sufficiently general for our purposes was proved by Kuranishi [1]. An important special case has been proved by Chevalley [2]. A generalization was proved by Iwasawa [1] and by Gleason [4].

A group or local group is of *class* C^k if it is locally euclidean and if coordinates can be chosen in a neighborhood of e such that the group multiplication is given by

$$f_i(x_1, \ldots, x_n; y_1, \ldots, y_n), \quad i = 1, \ldots, n$$

where each f_i is of class C^k. By the theorem on implicit functions the inverse is automatically of class C^k. For the same reason the analyticity of f_i, $i = 1, \ldots, n$ (Lie groups) implies that the inverse is analytic.

A group is differentiable of class C^∞ if there exist coordinates in which the product-functions f_i are *of class C^k for every k*.

THEOREM. *Let G be a locally compact group and let N be a closed central subgroup, locally an n-parameter vector group. If G/N is an m-parameter group of class C^k and if there exists a local cross-section K of the local left N-cosets, then G is an $(m + n)$-dimensional group of class C^k.*

Kuranishi's theorem is more general; in it N is not necessarily a central subgroup but is merely abelian and invariant.

Let G' denote G/N. By hypothesis, in a suitable neighborhood of e in G', the elements $s, t, \ldots$ are in one-one correspondence with sets of real coordinates (we follow Kuranishi [1])

1)　　$\alpha = (\alpha_1, \ldots, \alpha_m)$,　$|\alpha_i| < a$, for some $a > 0$,

in such a way that the product of elements of G'

$$s(\alpha)s(\alpha') = s(\alpha'')$$

is represented by functions f_i, $i = 1, \ldots, m$

2)　　$\alpha_i'' = f_i(\alpha_1, \ldots, \alpha_m; \alpha_1', \ldots, \alpha_m')$, $i = 1, \ldots, m$

whose partial derivatives exist and are continuous for all orders up to and including k. It follows too that the coordinates of $s^{-1}s'$ as well as of $s's^{-1}$ are of class C^k in the coordinates of s and s'.

The group N has local coordinates $x_1, \ldots, x_n$ and for $x, x' \in N$

3) $\quad x \cdot x' = (x_1 + x_1', \ldots, x_n + x_n')$
$\qquad x^{-1} = (-x_1, \ldots, -x_n);$

of course N is a local Lie group.

By hypothesis, also, there exists at e a local cross-section K, $e \in K$, of the local N-cosets. It follows that there is a homeomorphism ψ between a neighborhood of e in G' and a neighborhood of e in K:

4) $\quad \psi(s) \in K; \; s = s(\alpha) \in G'; \; |\alpha_i| < b$ for some $b > 0$.
$\qquad \psi(s(0, \ldots, 0)) = e \in K.$

In a sufficiently small neighborhood of e, every point of G is uniquely of the form

5) $\quad \psi(s) \cdot x = x \cdot \psi(s), \; x \in N, \; s \in G'.$

From the fact that K is a local cross-section at e it follows that there is a continuous function $u(s, t)$, s and t near e in G', described by coordinate functions $x_i(s, t)$, $i = 1, \ldots, n$:

6) $\quad u(s, t) = (x_1(s, t), \ldots, x_n(s, t)) \in N$

such that

7) $\quad \psi(s)\psi(t) = u(s, t) \cdot \psi(st).$

If we use the associative law in G

$$(\psi(s)\psi(t))\psi(r) = \psi(s)(\psi(t)\psi(r))$$

and remember 5) we obtain

$$u(s, t)u(st, r) = u(t, r)u(s, tr).$$

This implies for each coordinate function that

8) $\quad x_i(s, t) + x_i(st, r) = x_i(t, r) + x_i(s, tr), \; i = 1, \ldots, n.$

We now have a local coordinate system in G as follows:

$$(\alpha_1, \ldots, \alpha_m; x_1, \ldots, x_n).$$

The first m coordinates of the product of a pair of elements, $\psi(s) \cdot x$ and $\psi(t) \cdot y$, depend only upon the first m coordinates of the factors and are of class C^k according to 2); the last n depend upon all coordinates. Their dependence on the coordinates of s and t is given by the functions $x_i(s, t)$ which are continuous merely but which satisfy the functional equation 8). We must show that they can be replaced by comparable functions of class C^k. This proceeds in three stages.

Let $\lambda(s)$ be any non-negative real function on G' of class C^∞ in the coordinates 1), not identically zero, whose carrier is inside the coordinate neighborhood.

We introduce a left invariant measure in G' such that $\int \lambda = 1$, and define the following auxiliary functions

9) $\quad y_i(s) = \int_{G'} x_i(s, t)\lambda(t)\,dt, \; i = 1, \ldots, n;$

the symbol dt calls attention to the fact that we are integrating a function of t and that s is a parameter. For each t, the integrand varies continuously with s and since it has a compact carrier it is uniformly continuous. Therefore by property 3) of 2.17.3, $y_i(s)$ is continuous in s.

Let $v(s) = (y_1(s), \ldots, y_n(s)) \,\epsilon\, N$, and define $\psi'(s)$ by

10) $\quad \psi'(s) = v(s)^{-1}\psi(s).$

Then

$$K' = \{\psi'(s), \; s \text{ near } e \text{ in } G'\}$$

is a new local cross-section in G of N-cosets. Near e each element is uniquely of the form:

$$x'\psi'(s) = \psi'(s)x'; \; \psi'(s) \,\epsilon\, K', \; x' \,\epsilon\, N.$$

Let $u'(s, t) \,\epsilon\, N$ with coordinates

$$x_1'(s, t), \ldots, x_n'(s, t)$$

be defined by

11) $\quad \psi'(s)\psi'(t) = u'(s, t)\psi'(st).$

Putting 10) into 11), remembering always that N is central, we obtain

$$v(s)^{-1}v(t)^{-1}u(s, t)\psi(st) = u'(s, t)v(st)^{-1}\psi(st).$$

Cancelling $\psi(st)$ and expressing the relation in the coordinates of 3) we get:

12) $x_i'(s, t) = x_i(s, t) - y_i(s) - y_i(t) + y_i(st).$

Using 9) and the fact that $\int \lambda = 1$:

$$x_i'(s, t) = \int_{G'} [x_i(s, t) - x_i(s, r) - x_i(t, r) + x_i(st, r)]\lambda(r)dr.$$

Then using 8)

$$x_i'(s, t) = \int_{G'} [x_i(s, tr) - x_i(s, r)]\lambda(r)dr.$$

Using the left invariance of the integral in the first integrand we obtain:

13) $x_i'(s, t) = \int_{G'} x_i(s, r)\lambda(t^{-1}r)dr - \int_{G'} x_i(s, r)\lambda(r)dr.$

The function $\lambda(t^{-1}r)$ is of class C^k in the coordinates of $t \,\epsilon\, G'$ and therefore the first integrand in 13) is of class C^k in these coordinates; the second is independent of them. It follows as in the usual way using property 3) of 2.17.3 that the functions $x_i'(s, t)$ are of class C^k in the coordinates of t. This completes the first stage of the construction.

Let $\int^*$ denote a right invariant integral normalized by the condition that

14) $\int_{G'}^* \lambda = 1.$

The functions $x_i'(s, t)$ are in the same relation to the cross-section K' that the functions $x_i(s, t)$ were to the set K and they satisfy the functional equation 8). Substituting them into that equation, multiplying by $\lambda(s)$ and integrating over G' by the right invariant $\int^*$ we obtain.

15) $\int^* x_i'(s, t)\lambda(s)ds + \int^* x_i'(st, r)\lambda(s)ds$
      $= \int^* x_i'(t, r)\lambda(s)ds + \int^* x_i'(s, tr)\lambda(s)ds.$

The first term on the left has been shown to be of class C^k in the coordinates of t and this is also true of the last term on the right because of the nature of the dependence of tr on t. The integrand $x_i'(st, r)\lambda(s)$ regarded as a function $f(s)$ and subjected to a right translation:

$$f(s)t = f(st^{-1})$$

becomes

$$x_i'(s, r)\lambda(st^{-1}).$$

Therefore

$$\int\!\!{}^{*}x_i'(st, r)\lambda(s)ds = \int\!\!{}^{*}x_i'(s, r)\lambda(st^{-1})ds$$

and this term also is of class C^k in the coordinates of t. The remaining term in 15) is simply $x_i'(t, r)$ in view of 14). It follows that $x_i'(t, r)$ is of class C^k in the coordinates of t and from what was learned above $x_i'(t, r)$ is also of class C^k in the coordinates of r. This is the second stage of the proof.

Finally let us operate on the functions $x_i'(s, t)$ exactly as we did on $x_i(s, t)$, defining auxiliary functions analogous to 9)

$$y_i'(s) = \int x_i'(s, t)\lambda(t)dt, \quad i = 1, \ldots, n,$$

using the original left invariant integral. We obtain

16) $x_i''(s, t) = \int x_i'(s, r)\lambda(t^{-1}r)dr - \int x_i'(s, r)\lambda(r)dr.$

In the light of the remarks following 13) relation 16) shows that the new functions $x_i''(s, t)$ are of class C^k in s and t simultaneously. Analogously to 10) we define a function $v'(s) \, \epsilon \, N$, $s \, \epsilon \, G'$ and a map $\psi''(s) \, \epsilon \, K''$, a new local section. In a neighborhood of e, elements of G are uniquely represented in the form

$$p = x\psi''(s) = \psi''(s)x.$$

The coordinates of p in G are

$$(\alpha_1, \ldots, \alpha_m; x_1, \ldots, x_n),$$

the first m being coordinates of s in G' and the last n being coordinates of x in N.

If $p = \psi''(s)x$ and $q = \psi''(t)y$ then

17) $pq = \psi''(st)u''(s, t)xy,$

with u'' analogous to u' in 11). As once before, the first m coordinates of the product pq are of class C^k in the first m coordinates of p and q. The last n coordinates are

$$x_1''(s, t) + x_1 + y_1, \ldots, x_n''(s, t) + x_n + y_n.$$

By what has been shown above these are of class C^k in the coordinates of s and t, and they are obviously linear in the remaining coordinates. This completes the proof.

In the case which interests us most G/N is a Lie group with an analytic coordinate system. Such a system is of class C^∞. Then, since the function λ was chosen to be of class C^∞ the argument above gives a coordinate system in G which is of class C^k for every k. Therefore the group G becomes a differentiable group of class C^∞. However, the new coordinate system is not a canonical one, in general. The fact that any canonical coordinate system is automatically of class C^∞, and even analytic is difficult to prove.

4.4. Analytic coordinates

In a differentiable group of class C^k, $k \geq 1$ it is always possible to introduce a canonical coordinate system by a transformation of coordinates of class C^k, and the group operations become analytic in these new coordinates. This was first proved in 1891 by F. Schmidt (for $k = 2$) and is the result referred to by Hilbert in the quotation in 2.15. A detailed presentation of this subject along somewhat the same lines as that of Schmidt will be found in Eisenhart [1] and also in Pontrjagin [1]. The theorem was extended to a much wider class of groups by the work of G. Birkhoff [1], I. E. Segal [1] and P. Smith [4]. We shall not give the details of this classic theorem but shall sketch some of the ideas in it at the end of the section. In view of this theorem, and of what we have already proved up to this point we may now state the following principal theorem.

THEOREM. *A locally compact connected group G without small subgroups is a Lie group.*

If U_0 is a neighborhood of e, U_0^s without subgroups and L is the set of points in U_0 which lie on one-parameter local groups radiating to the boundary of U_0, then L is locally homeomorphic to a vector-space of some dimension $n \geqq 1$.

In the vector-addition $x \circ y = \lim (x(1/m)y(1/m))^m$, associated with one-parameter subgroups x and y of L, L is locally isomorphic to the n-dimensional vector space.

If $x_1, \ldots, x_n$ denote n one-parameter local groups in L which are linearly independent in the vector-addition in L then every element of L near enough to e is uniquely represented as

$$a_1 x_1 + \ldots + a_n x_n$$

for some choice of real numbers $a_1, \ldots, a_n$. If these numbers are taken as coordinates of elements near e, then the group operations are analytic functions of these coordinates.

We now give a brief survey of some of the ideas in the proof of the fact that a differentiable group is a Lie group. Let G be a differentiable group of class C^k, $k \geqq 3$. The elements x, y, z, etc. near e are given by n real coordinates, x_i, y_i, z_i, etc., with $e = (0, \ldots, 0)$. If $z = xy$, then

$$z_i = f_i(x_1, \ldots, x_n; y_1, \ldots, y_n)$$

and the functions f_i are of class C^k. They can be expanded using Taylor's Theorem which for $k \geqq 3$ has the form

$$f_i = x_i + y_i + \Sigma_{j,k} a_{ijk} x_j y_k + R_i.$$

The numbers a_{ijk} depend on G and on the coordinate system. It turns out that the n^3 numbers

$$c_{ijk} = a_{ijk} - a_{ikj}$$

are of great importance. They are called the *structural constants*, or the *constants of composition* of the group. They satisfy a "relation of Jacobi" in addition to the obvious skew-symmetry in j and k; any set of such constants determines a local Lie group, unique to within a local isomorphism.

The partial derivatives $\partial f_i / \partial y_j = \delta_{ij}$ at $x = e$. However, these partials at $y = e$ are functions of x depending on the coordinate system.   Let

$$v_{ij}(x) = v_{ij}(x_1, \ldots, x_n) = \frac{\partial f_i}{\partial y_j}(x_1, \ldots, x_n; 0, \ldots, 0).$$

The importance of the functions v_{ij} is indicated by the following theorem.   Let $a_1, \ldots, a_n$ be real constants, not all zero; they define a "direction" at e. Let $x_i(t)$ be functions of a parameter t satisfying the system of total differential equations:

$$dx_i(t)/dt = \Sigma_j a_j v_{ij}(x_1, \ldots, x_n)$$
$$x_i(0) = 0.$$

Then the set of elements of G: $x(t) = (x_1(t), \ldots, x_n(t))$ is a one-parameter local group (for sufficiently small t).   Conversely, the one-parameter subgroups of G satisfy such equations, and the direction $(a_1, \ldots, a_n)$ is the tangent vector at e.   Because of this theorem it is possible to introduce into G a canonical coordinate system of class C^k.

In canonical coordinates the one-parameter groups are sets of elements with coordinates of the form $x_i = a_i t$, where (a_i) is an "initial direction".   From this point of view the functions v_{ij} yield functions $w_{ij}(t; a)$ $tv_{ij}(at)$.   The v_{ij} satisfy certain partial differential equations involving the structural constants.   When canonical coordinates are used these equations become equivalent to the system of total differential equations:

$$\frac{dw_{ij}}{dt} = \delta_{ij} + \Sigma_{j,k} c_{ijk} a_j w_{ik}.$$

This shows that in canonical coordinates the structural constants determine the functions v_{ij} (through the w's) as unique analytic functions of the coordinates $a_i t$ (see Pontrjagin, loc. cit.).

The product-functions $z_i = f_i(x; y)$ satisfy a certain system of partial differential equations, namely:

$$\Sigma_j v_{ij}(z) \frac{\partial z_j}{\partial x_k} = v_{ik}(x); \quad i, k = 1, \ldots, n,$$

which are called the *fundamental differential equations of the group*.

These equations determine the functions z_i if the v's are known, and if certain "integrability conditions" are satisfied. These conditions link the functions v_{ij} and their derivatives to the structural constants. The v's having been determined so that the integrability conditions shall be satisfied, it follows that the functions z_i are analytic if the v's are analytic. This concludes the summary.

4.5. Small subgroups

It will be remembered that a locally compact group may have arbitrarily small subgroups. For example, the product (1.6) of an infinite number of compact groups has small subgroups of the same product-structure as itself. All totally-disconnected groups have small open subgroups. The solenoidal groups (2.14) are one-dimensional non-Lie groups which have arbitrarily small totally disconnected subgroups. We shall see in 4.9 that small subgroups of connected locally compact finite-dimensional groups are totally disconnected and belong to the center. In the case of connected locally compact infinite-dimensional groups small subgroups are not necessarily invariant.

In this section we shall prove that the sufficiently small subgroups of a locally compact group G themselves generate a small group; if G/G_0 is compact we can always find small invariant subgroups which are generated in this way. This result was first proved in complete generality by Yamabe [4], the key idea being the Lemma in 3.3 which connects the powers A^n of "small" subsets A of G and the powers of individual elements contained in A.

LEMMA. *Let G be a locally compact group and let U be a neighborhood of e. There is a subgroup H contained in U and there is a neighborhood W of e such that every group contained in W is a subgroup of H.*

COROLLARY. *There is an open subgroup G' in G and an invariant (in G') compact group H' in $U \cap G'$ such that G'/H' has no small groups.*

First take the case that G has a countable base at e. Let W_n, $n \in I$, be a monotonic sequence of compact symmetric neighborhoods of e intersecting in e. For convenience, let $W_1 \subset U$. For each n, let A_n denote the set of elements of W_n all of whose powers are in W_n:

$$A_n = \{x;\ x, x^2, \ldots \in W_n\}.$$

Then for every $V(e)$, almost all sets A_n are contained in V and all powers of individual elements of these sets remain inside of V. This contradicts the conclusion of Lemma 3.3. Accordingly, we shall show that there is an integer n' such that for all k

1) $A_{n'}^k \subset W_1 \subset U$.

If this were not true there would be a sequence of integers $k(n)$ such that

2) $A_n^{k(n)+1} \not\subset W_1,\ A_n^{k(n)} \subset W_1.$

Without loss of generality, we may suppose that a set E exists defined by

3) $\lim A_n^{k(n)} = E \subset W_1,$

and we know that E meets the boundary of W_1 (3.1).

This contradicts Lemma 3.3 *unless* $E = E^2$. If $E = E^2$ then E is a compact group (3.1). Using what we know (2.16) about the structure of compact groups we shall obtain a contradiction in this case too. There is an invariant subgroup E' of E contained in the interior of U such that E/E' has no small subgroups. As we have already observed in 2.16 this implies that there is some compact neighborhood U' of E' such that every subgroup of E in U' is also in E'. Of course in 2.16 the set U' would correspond to a subset of E and E' would be a set of interior points of U' relative to E. But in the present case we may suppose that U' is in W_1 and that E' consists of inner points of U' (using the relative topology). We shall now get a contradiction as before, using the same sequence of sets A_n and the compact set U' instead of the previous W_1.

Since U' is in W_1, it follows from 2) that there is a sequence of integers $k'(n) \leq k(n)$ such that $A_n^{k'(n)+1} \not\subset U'$, but $A_n^{k'(n)} \subset U'$. We may suppose that there is a set F defined by $\lim A_n^{k'(n)}$ which reaches to the boundary of U'. Now, finally, if $F = F^2$, then it is a group and since it is a subgroup of E and belongs to U' it follows by construction that it belongs to E'. In this case of course it cannot reach the boundary of U', and this is a contradiction. On the other hand if we do not have $F \neq F^2$, then Lemma 3.3 is contradicted exactly as in the first paragraph of the proof.

We have now shown that there is an integer n' such that 1) holds. Let H denote the closed group generated by $A_{n'}$, i.e. H is the closure of the union of all powers of $A_{n'}$. Then H is the group we are seeking and $W_{n'}$ is the neighborhood W called for in the Lemma.

We turn now to the proof of the corollary, using the group H and the neighborhood W which have just been constructed. Since H is compact, it has an invariant compact subgroup H' consisting of interior points of W such that H/H' has no small subgroups. There is an open set W', $H' \subset W' \subset W$ such that every subgroup of G which belongs to W' is also a subgroup of H'.

By 2.1, since H' is compact and W' is open there exists a compact symmetric neighborhood V' of e such that

$$V'H'V' \subset W',$$

and in particular such that

$$g^{-1}H'g \subset W', \quad g \epsilon V'.$$

Let G' be the *normalizer* of H'; i.e. G' is the group of elements g such that $g^{-1}H'g = H'$. We shall show that G' is open by showing that it contains V'.

If a group belongs to W' it belongs to W, and therefore belongs to H. On the other hand if a group belongs to H and to W', then it belongs to H'. This shows that if $g \epsilon V'$, the group $g^{-1}H'g$ belongs to H'. But since V' is symmetric, we see that for g in V', also $gH'g^{-1}$ belongs to H'. From this it follows that g is in G'.

In case G is not metric, i.e. in case G does not have a countable base at e, the Lemma and Corollary are still true. We shall need a similar argument in the next section and show the details there.

4.6. Main approximation theorem

THEOREM. *Let G be a locally compact group such that G/G_0 is compact, G_0 being the identity component. Let U be an arbitrary neighborhood of e. There exists in U a compact invariant subgroup H such that G/H has no small subgroups; and G/H is isomorphic to a Lie group.*

It is convenient to take the metric case first. By the preceding section, there is an open subgroup G' of G and an invariant (in G') subgroup $H' \subset G'$ such that G'/H' has no small subgroups. There is also by 4.5 an open set W' containing H' such that any subgroup in W' is a subgroup of H'.

Since G' is open and closed, G' contains G_0. The natural map of G onto G/G_0 carries $G' \subset G$ onto the open subset G'/G_0. It follows that there exist a finite number of translations of G':

$$g_1 G', \ g_2 G', \ldots, g_n G', \ g_i \in G$$

whose union covers G.

Let

$$H = \cap \ g_i H' g_i^{-1}, \ i = 1, \ldots, n,$$

and let

$$W = \cap \ g_i W' g_i^{-1}, \ i = 1, \ldots, n.$$

The group H is invariant in G. This is becauses H' is invariant in G' and therefore:

$$H = \cap \ (g_i g) H' (g_i g)^{-1}, \ i = 1, \ldots, n, \ g \in G'.$$

But then

$$H = \cap \ f^{-1} H' f, \ f \in G$$

since every element of G is in the form $g_i g$ for some g_i and some g of G'. This shows that H is invariant in G.

Any subgroup in W is contained in each $g_i^{-1} W' g_i$ and therefore is a subgroup of each $g_i^{-1} H' g_i$; then it is also a subgroup of H.

Since H is contained in W this shows that G/H has no small subgroups. Therefore G/H is isomorphic to a Lie group, and the theorem is proved in the case that G is metric.

In the general case there is a compact invariant subgroup H^* contained in the interior of U such that G/H^* is metric (2.6). Let

$$T : G \to G/H^*.$$

Then TU is a neighborhood of e in G/H^*. Suppose for a moment that TU contains an arbitrarily small invariant subgroup H' such that $(TG)/H'$ is a Lie group. Let $H = T^{-1}H' \subset G$. Then H is the desired group such that G/H is a Lie group and H belongs to U provided H' is chosen sufficiently small.

We shall now see that such an H' exists. We know

$$TG_0 \subset (G/H^*)_0$$

so if $T^{-1}(G/H^*)_0 = F$

$$G_0 \subset T^{-1}(G/H^*)_0 = F$$

and F is a closed invariant subgroup of G. Now

$$G/G_0 \to G/F = (G/H^*)/(G/H^*)_0$$

and since G/G_0 is compact it follows that $(G/H^*)/(G/H^*)_0$ is compact. This proves the existence of H' as desired, which proves the Theorem. This also completes the proof of the Lemma and Corollary of the last section.

4.7. *Inverse sequences*

We have not yet defined the topological dimension of a locally compact group. We could begin by saying that any open set in a euclidean space $E = R_1 \times \ldots \times R_n$ has dimension n. It was proved by Brouwer that this definition of the dimension of such open sets is invariant, that is that an open set in E can be homeomorphic to an open set in $E' = R_1 \times \ldots \times R_m$ only if $m = n$. The dimension of a Lie group G can be defined as the dimension of any open set in a euclidean space which is homeomorphic to an open subset of G.

For a general locally compact metric group G, the dimension can be defined as is done for any general space in dimension theory (see Hurewicz-Wallman [1]). For such groups G, however, the dimension can also be defined as the maximum dimension of any cell contained in G if such a maximum exists, otherwise as infinity. We shall point out later that these two definitions of dimension are equivalent for groups.

The following two sections will be useful in considering dimension.

4.7.1. LEMMA. *Let G be a locally compact group with compact invariant subgroups H and F. If G/H and G/F have no small subgroup then $G/(H \cap F)$ has no small subgroups.*

In G there are open sets U and V

$$F \subset U, \ H \subset V$$

such that if a subgroup K of G is in U then it is in F and if it is in V then it is in H. Let T be the map

$$T : G \to G/(H \cap F)$$

and let W' be a neighborhood of e in $G/(H \cap F)$ such that

$$T^{-1}(W') \subset U \cap V$$

and let K' be a subgroup of $G/(H \cap F)$, $K' \subset W'$. Then

$$T^{-1}(K') \subset U \cap V$$

and

$$T^{-1}(K') \subset F, \ T^{-1}(K') \subset H.$$

Therefore

$$T^{-1}(K') \subset F \cap H$$
$$K' = e,$$

as was to be proved. It follows that if G/F does not have small groups then F includes a group F_1 where F_1 is arbitrarily small invariant in G (G/G_0 compact) and G/F_1 has no small subgroups.

4.7.2. Let G be metric and let G/G_0 be a compact group. Then it follows from the Lemma and from 4.6 that one can find a sequence of compact invariant subgroups of G:

$$H_1 \supset H_2 \supset \ldots$$

intersecting finally in e such that the corresponding factor groups and the associated natural maps:

$$G/H_1 \xleftarrow{\;T_1\;} G/H_2 \xleftarrow{\;T_2\;} \ldots \xleftarrow{\quad} G/H_n \xleftarrow{\;T_n\;} \ldots$$

is an inverse projection sequence (2.7) of groups without small subgroups and the limit group is G.

We now amplify some of the earlier remarks (2.7) about such limit groups. Let P_n, for each n, denote the natural map

$$P_n : G \to G/H_n.$$

If the sequence $x_n \in G/H_n$ forms an inverse sequence, that is if for every n:

$$T_n(x_{n+1}) = x_n,$$

then the following inclusions hold for the indicated subsets of G:

$$P_{n+1}^{-1}(x_{n+1}) \subset P_n^{-1}(x_n).$$

Furthermore the set

$$\cap P_n^{-1}(x_n)$$

is a unique point of G, depending of course on the sequence $x_1 \leftarrow x_2 \leftarrow \ldots$.

Suppose now that $x_n(t) \subset G/H_n$ is an inverse sequence of one-parameter groups, so parameterized that for each value t'

$$T_n x_{n+1}(t') = x_n(t').$$

Then the reader will have no difficulty in showing that the sequence

$$x_1(t) \leftarrow x_2(t) \leftarrow \ldots$$

defines a one-parameter group $x(t) \subset G$, and for every t'

$$P_n x(t') = x_n(t').$$

Suppose, next, that we are given a one-parameter subgroup

$x_1(t) \subset G/H_1$. We shall now find a "covering" group $x_2(t) \subset G/H_2$. Let U_2 be a neighborhood of the identity of G/H_2 uniquely ruled by one-parameter groups and let U_1 be a similar neighborhood of the identity of G/H_1 such that

$$U_1 \subset T_1 U_2.$$

There is a parameter-value t' such that $x_1(t') \, \varepsilon \, U_1$ and there is an element $x_2 \, \varepsilon \, U_2$ such that $T_1 x_2 = x_1(t')$ and there is a unique one-parameter group

$$x_2(t) \subset G/H_2$$

determined by x_2. If we normalize $x_2(t)$ by requiring that $x_2 = x_2(t')$, for the same t' as above, then we shall have

$$T_1 x_2(t) = x_1(t)$$

for all t.

The reader will have no difficulty in seeing how to continue inductively and in this way construct an inverse sequence,

$$x_1(t) \leftarrow x_2(t) \leftarrow \ldots$$

defining a limit one-parameter group $x(t) \subset G$.

4.8. Existence of n-cells in a group

THEOREM. *Let G be a locally compact metric group and let H_1 be a compact invariant subgroup such that G/H_1 is an N-dimensional group without small subgroups. Then G contains an N-cell.*

Let $u_1(t_1), \ldots, u_N(t_N) \subset G/H_1$ denote local one-parameter groups, $|t_i| \leq a$, $a > 0$, which are independent in a vector addition (3.9.9) associated with G/H_1. We suppose further, that the following set K_1 is defined and associative in this addition:

$$K_1 = \{u_1(t_1) \circ \ldots \circ u_N(t_N); \ |t_i| \leq a\}.$$

Now let $x_1(t), \ldots, x_N(t) \subset G$, $|t| \leq a$, be local one-parameter groups which are constructed for us by 4.7.2. Then for each $i = 1, \ldots, N$ and each parameter-value $t' \, \varepsilon \, [-a, a]$:

$$P_1 x_i(t') = u_i(t').$$

Next, let n be a momentarily fixed integer and consider the set of local one-parameter groups $v_i(t) = P_n x_i(t) \subset G/H_n$, $i \leq N$.

Before we speak of a vector addition in G/H_n we recall that the constructions in 3.9 rest on a choice of a sequence of integers. But, given such a sequence, say I_1, for G/H_1, we can use some *subsequence* $I_n \subset I_1$ to define a vector addition in G/H_n. On this understanding, there is a real number $b > 0$ such that the following set K'_n is defined and associative in a vector addition in G/H_n:

$$K'_n = \{v_1(t_1) \circ \ldots \circ v_N(t_N); \; |t_i| \leq b\}.$$

We shall restrict b by the condition $b \leq a$.

Let Q_n denote the natural map of G/H_n onto G/H_1. It is clear that these maps are related to the maps P_n through

$$Q_n P_n x = P_1 x, \quad x \, \varepsilon \, G.$$

For each $i = 1, \ldots, N$, and each parameter value t'

$$Q_n v_i(t') = u_i(t').$$

The reader will have no difficulty in showing that Q_n maps K'_n isomorphically (in the respective vector-additions) into K_1 and in fact into the subset K'_1:

$$K'_1 = \{u_1(t_1) \circ \ldots \circ u_N(t_N); \; |t_i| \leq b\}.$$

We want an N-cell $K_n \subset G/H_n$ which shall exactly cover K_1, and this is now easy to get. Each local one-parameter group $v(t) \subset K'_n$ is part of a one-parameter group defined for all values of the parameter and we can choose as large an interval as we may need. We can do this systematically by noticing that the larger interval is to the original one in the fixed ratio a/b.

This gives us the desired set $K_n \subset G/H_n$. It follows that

$$Q_n K_n = K_1.$$

The map Q_n is one-one on K_n to K_1 because if two distinct points of K_n could map into the same point of K_1 it would follow that two distinct one-parameter groups of K_1 had a point in common. But this is in contradiction to the choice of K_1.

Now it is clear how we can pick an N-cell $K_n \subset G/H_n$ for each n, covering K_1 and the reader can show that in this way we get an inverse sequence

$$K_1 \leftarrow K_2 \leftarrow \ldots$$

defining an N-cell in G.   See also 4.15.

4.8.1.   Suppose that G_1 and G_2 are groups with no small subgroups, and suppose that we are given a homomorphism Q of G_2 upon G_1. Then there is a vector addition in G_2 and a corresponding vector addition in G_1 such that the associated vector-spaces M_1 and M_2 (as in 3.10) are homomorphic; and this homomorphism $Q*$:

$$Q*M_2 = M_1$$

is defined as follows. Let T_i denote the exponential map (3.10) of M_i into G_i, $i = 1, 2$. Now given $v \, \varepsilon \, M_2$, there is one and only one $u \, \varepsilon \, M_1$ such that

$$T_1 Q*v = Q T_2 v = T_1 u \, \varepsilon \, G_1.$$

It follows from this that if $H \subset G_2$ is the kernel of the homomorphism Q and $K \subset M_2$ is the kernel of the homomorphism $Q*$ then

$$T_2 K \subset H$$

and T_2 is the exponential map on the vector space K to the group H.

4.9. Finite-dimensional metric groups

Compact spaces are known which are infinite-dimensional in the topological sense and which do not contain any subset homeomorphic to a one-cell. In the case of locally compact groups, however, the topological dimension is precisely the upper bound of the dimension N of subspaces homeomorphic to N-cells. This upper bound may be infinite as the inifinite-dimensional torus (1.11.2) shows. In this section we shall characterize the finite-dimensional metric groups. The supporting material necessary for the non-metric groups will be found in 4.15.

THEOREM. *Let G be locally compact, metric, and suppose that G/G_0 is compact. Suppose that every approximating group without small subgroups is at most N-dimensional for a fixed integer N. Then G is locally the product of a compact totally disconnected group Z and a local m-parameter Lie group R. The group G cannot have arbitrarily small connected subgroups.*

Conversely, if G does not have arbitrarily small connected subgroups then, for some m, G has the local product form announced above and every approximating group without small subgroups is of dimension $\leqq m$.

Suppose that N is the maximum integer for which there is a compact invariant subgroup H_1 of G such that G/H_1 is N-dimensional and is without small subgroups. Now let $H_2 \subset H_1$ be any compact invariant subgroup such that G/H_2 is without small subgroups. Then $G_2 = G/H_2$ is N-dimensional, by the choice of N, and the natural map Q_1 of G_2 upon $G_1 = G/H_1$, has a kernel which is zero-dimensional (see 4.8.1). This kernel is isomorphic to H_1/H_2. Since we are dealing with Lie groups the zero-dimensional kernel must be discrete and it follows that H_2 is open in H_1. Since H_1/H_2 is compact, H_1/H_2 must also be a finite group.

We see now that H_1 has arbitrarily small open (and closed) invariant subgroups and therefore H_1 is totally disconnected. Of course, every subgroup of H_1 is also totally disconnected. Since G/H_1 has no small subgroups there is a neighborhood U_1 of H_1 such that every subgroup of G which is in U_1 is also in H_1. Then it follows that G does not have small connected subgroups. There remains to find the local Lie group R and to prove that G has the desired product form.

4.9.1.   Let $K_1 \subset G/H_1$ be an N-cell neighborhood of the identity of G/H_1 which is uniquely ruled by one-parameter groups and which is sufficiently small so that we can "raise" it, by the method of 4.8 to an N-cell $K \subset G$. The set K_1 is necessarily a local group (being a neighborhood of the identity of a group) and all the sets K_n which are constructed by the method of 4.8 are local groups in the present case, for the same reason. The sequence

$$K_1 \leftarrow K_2 \leftarrow \ldots$$

is an inverse projection sequence and it follows that the set K is also a local group; for example, it is easy to see that for elements of K near to e the product element is also in K.

Then we can find an N-cell neighborhood $R \subset K$ of e such that $RR^{-1} \subset K$; R is a local group. Because the map of G onto G/H_1 is one-one on K, it is clear that R and also K is isomorphic to a local Lie group. Observe next that every element of K commutes with every element of H_1. *For*, the set

$$\{k^{-1}hk; \ k \in K, \ h \text{ fixed in } H_1\}$$

is connected because K is connected and it belongs to H_1 because H_1 is invariant. Therefore this set reduces to the element h because H_1 is totally disconnected and each element is a component.

The set $K \cap H_1 = e$ because under the map P_1 of G onto G/H_1 every element of H_1 maps into the identity of G/H_1 but no element of K does this (excepting e). Of course, then

$$R^2 \cap H_1 = e$$

by the choice of R. Now we can see that the elements of the product set $RH_1 \subset G$ cover a neighborhood of e in a unique way.

First, by the fact that P_1 is one-one on K to G/H_1 it follows that P_1R is a neighborhood of the identity in G/H_1 and therefore RH_1 which is $P_1^{-1}(P_1R)$ is a neighborhood of e in G. If it were possible to find elements x, y in R and h, h' in H_1 such that

$$xh = yh',$$

then it would follow that

$$x^{-1}y \in H_1 \cap K.$$

This means that $x = y$ and then it is clear that furthermore $h = h'$.

Finally, since every element of R has been seen to commute with every element of H_1 it is clear that RH_1 is a group product as well as a space product and this completes the proof of the direct part of the theorem.

4.9.2. As for the converse part of the theorem we can see at once that if there is a neighborhood $U \subset G$ of the identity which does not have any connected subgroups and if $H_1 \subset U$ is such that G/H_1 has no small subgroups, then H_1 is totally disconnected. Now any $H_2 \subset H_1$ is totally disconnected. If G/H_2 does not have small subgroups then it follows that also H_1/H_2 is without small subgroups and therefore H_1/H_2 is a finite group. Therefore G/H_2 is of the same dimension as G/H_1. Now the methods of 4.9.1 apply to this case also and the theorem may be considered to be completely proved.

4.9.3. It is a theorem of dimension-theory that a compact set is at most n-dimensional if for every preassigned $\epsilon > 0$ it can be mapped upon an n-dimensional polyhedron by a continuous map $f(x)$ such that all the inverse sets $f^{-1}(f(x))$ are of diameter less than ϵ. A set is of dimension at least n if it contains a subset of dimension n. A euclidean n-cell is of dimension n. It follows from these theorems that *a locally compact metric group is n-dimensional in the topological sense if and only if it is the local product of an n-parameter local Lie group and a compact totally disconnected group.*

We showed before that if a group G is *not* finite-dimensional then it contains euclidean n-cells of arbitrarily large dimension. However a stronger result is true. It follows from a theorem of Iwasawa [1] which we shall not prove here that in this case (if G is connected) then G is locally the product of an arbitrarily small compact invariant subgroup and a local Lie group of arbitrarily large dimension.

4.10. The Hilbert Problem

Hilbert's question (2.15) about the analytic character of locally euclidean groups is now answered by the following theorem.

THEOREM. *A locally euclidean group has no small subgroups and is isomorphic to a Lie group.*

Let G be of dimension n. Then G is locally the product of an n-parameter local Lie group L and a compact totally disconnected group N. If K denotes a sufficiently small connected set in G containing e then K belongs to L. Therefore sufficiently small neighborhoods of e in G belong to L. Therefore L is a neighborhood of e in G and N is a discrete group or N reduces to e. This completes the proof.

4.10.1.   The same argument proves the following more general characterization of a Lie group.

THEOREM.   *A locally compact group which is finite-dimensional and locally-connected is a Lie group.*

A space is *locally-connected* if each point has small connected neighborhoods.

The condition of finite-dimensionality in the Theorem is essential; the infinite-dimensional toral groups (1.6), for example, are locally connected.

4.10.2.   A certain part of the topology of finite-dimensional groups was already known through the work of Montgomery [7, 8, 9, 10, 11], Gleason [4], Iwasawa [1] and others, before the theorems of this and the preceding section had been proved. Thus in Montgomery-Zippin [13] on the theorems of 4.9, 4.10 we were able to depend *from the beginning* on the finite-dimensionality of the group G and to proceed by dimension-theoretic methods. This work is not easily summarized, and the interested reader is referred to the paper. However, there is one aspect of the work which it is worth calling attention to here, although it will come up again in other connections.

Suppose for the moment that G denotes a finite-dimensional locally compact space which has many of the properties of a manifold. Suppose that Z is a compact totally disconnected group which is given as an effective transformation group of G. It has often been conjectured but it has not been proved that in this case Z must be a finite group. The problem is unsolved even when G is the euclidean 3-space. If it is *given* that Z is a finite group, then a theorem of Newman [1] asserts that orbits of Z cannot be uniformly small. This indicates a possible approach to the question of existence of small subgroups (at least in groups like locally-euclidean groups) and, in fact, it was this connection which inspired Newman's theorem. The connection, of course, is in the fact that a subgroup of a group G is a transformation group of G in certain natural ways.

If G is a group and Z a subgroup then to each $z \, \epsilon \, Z$ there is associated the transformation of G

$$T : g \to zg.$$

Now, if $z \, \epsilon \, Z$ is of finite order n, that is if $z^n = e$, then T^n is the identity map, and Newman's theorem implies that if G has appropriate properties then at least one of the powers of z lies outside of a certain fixed neighborhood of e. Thus it comes about that G with appropriate properties cannot have arbitrarily small *finite* groups.

Again, to each $z \, \epsilon \, Z$ one may associate the transformation

$$T : g \to z^{-1}gz.$$

Suppose now that it is discovered in some way that a subgroup Z' of Z is completely ineffective, — then one may conclude that Z' belongs to the center of G. The first situation described is the one which is appropriate to subgroups of a locally-euclidean group, the second is appropriate to a study of the small subgroups of a finite-dimensional group. The methods which the authors use in the paper cited applies both of these ideas.

Those methods, as they stand, are largely restricted to the case of a subgroup acting on a group in one of the ways indicated. But it is hoped the paper may be useful in the problem of arbitrarily small transformation groups; perhaps this will be the case with the argument in section 5, pp. 218—226, loc. cit.

4.11. Subgroups and factor groups of Lie groups

If G is a Lie group and H is a closed subgroup then H is a Lie group and the coset-space G/H is a manifold; if H is an invariant subgroup then G/H is a Lie group. This was proved by E. Cartan [1] generalizing the theorem on matrix groups (2.16); we point out in the next paragraph how to prove these facts from the results above.

If G is a locally euclidean group then G does not have small sub-groups, as one sees by the preceding theorem, and therefore the subgroup H does not have small subgroups and H is isomorphic to a Lie group. The introduction of a canonical coordinate system in G furnishes a vector addition giving rise to a pair of complementary linear vector-spaces, one of them spanning the group H and the other forming a local cross-section for the cosets of G/H (quite as in 2.16). In this way it is clear if H is invariant that G/H has no small subgroups.

4.12. Abelian groups

We saw in 2.21.1 that a locally compact abelian group G has an open subgroup G' generated by a compact subset, and G' has a compact subgroup H' such that G'/H' is an r-dimensional real vector group V_r. Now we shall prove the theorem of Pontrjagin [1] *that G' has a vector subgroup E_r and G' is the direct sum of H' and E_r.*

For non-metric groups see 4.15.

If we apply to G' the "lifting" process described in 4.7, 4.8 we obtain a local vector group L_r such that the map of G' onto G'/H' is a local isomorphism of L_r onto a neighborhood of the zero of V_r. Each local one-parameter subgroup of L_r is contained in a one-parameter *global* group $x(t)$ in G' and this group is mapped homomorphically into V_r. It follows from the Lemma in 2.21 that $x(t)$ is mapped isomorphically into V_r or else it is mapped into some compact subgroup. But V_r has no compact subgroups, except e, and it follows that the map is an isomorphism.

In this way, we see that L_r extends to a uniquely determined group E_r which is mapped isomorphically onto V_r; and E_r is an r-dimensional vector subgroup of G'. Clearly $H' \cap E_r = e$. Since each coset of G'/H' meets E_r in one and only one point, G' is the direct sum of H' and E_r as asserted.

COROLLARY. *The group E_r is also a direct summand of G.*

The proof is by induction on the number of cosets of G/G'. Let $a_1, a_2, \ldots,$ be an enumeration of elements of G *one from each coset of G/G'* in a finite or transfinite sequence, as the case may be. Let a denote the element a_1 *if a_1 is not in G'*, otherwise let a denote a_2. There are two possibilities to consider:

i) No power of a *belongs to G'*. In this case take the group generated by H' and a; denote it by H_1. Every element of H_1 is of the form ha^m for some element h of H' and some integer m. It is easy to see that H_1 and E_r have only the element e in common. Let G_1 be the group generated by H_1 and E_r; G_1 contains G' and a, and has E_r as direct summand.

ii) There is a least positive integer m such that $a^m = hv'$ for some h in H' and some v' in E_r. Then $a^m = hv^m$ for a uniquely determined v in E_r. Now let $b = av^{-1}$, and let H_1 denote the group generated by H' and b. As before, H_1 and E_r have only e in common and the group G_1 generated by H_1 and E_r has E_r as direct summand; G_1 contains G' and a_1.

Next, let a' denote the *first* element in the enumeration: $a_2, a_3, \ldots,$ which does *not* belong to G_1 and repeat the preceding construction. Thus one obtains a group G_2 which has E_r as direct summand and which contains G' and a_1 and a_2. Continuing this indicated process as long as may be necessary one actually constructs in this way a group H_* such that G is the direct sum of H_* and E_r. The group H_* contains H' and H_*/H' is discrete.

4.13. Two structure theorems

There are many classes of topological theorems about locally compact groups which are outside of the confines of this book; for some of these the only known proofs depend on the use of the Lie algebra theory. A very interesting survey of this field will be found in Samelson [2]. We shall call attention without proof to two theorems, giving some examples to illustrate them.

The first is a generalization by Iwasawa [1] of a theorem of Cartan [1]:

THEOREM. *A connected locally compact group G has maximal* *compact subgroups, and all such subgroups are connected and are conjugate to each other. Let K denote one of them. Then G contains subgroups $H_1, H_2, \ldots, H_r$ all isomorphic to the vector group V_1 and such that any element $g \epsilon G$ can be decomposed uniquely and continuously in the form*

$$g = h_1 \ldots h_r k, \quad h_i \epsilon H_i, \quad k \epsilon K.$$

In particular, the space of G is the product of the compact space of K and that of $H_1 \times \ldots \times H_r$, which is homeomorphic to the r-dimensional euclidean space E_r. The integer r may be called the characteristic index of G.

In the abelian case, the product $H_1 \ldots H_r$ is necessarily a group, and in the present set-up it would be a vector group so that this theorem generalizes the theorem of Pontrjagin in 2.21. A generalization to the *solvable* case was given by Malcev [1] and Chevalley [2].

To consider only simple examples of this theorem, the spaces associated with three-dimensional Lie groups are homeomorphic to one of the following spaces: the product of a line and a torus, the product of a plane and a circle, ordinary three-space. There is only one group corresponding to the product of a line by a torus, and this is the abelian group. There are several distinct instances of each of the other spaces (Eisenhart [1]). Thus the three-space is represented by the abelian group and also by the following, let us call it G_3 : G_3 is the subgroup of the 3×3 real matrices which has *one's* down the diagonal, *zero's* below it. We shall discuss this group in the next section.

The group of the projective transformations of the line is an example whose manifold is the product of a plane and a circle. Let us denote this group by P_3 : P_3 may be represented as the group of 2×2 real matrices of determinant *one*. This connected group has the interesting property that *not* every element lies on a one-parameter subgroup. For example, the element with *zero* in the lower left corner and $- 1$ elsewhere is not on a one-parameter subgroup of P_3; this follows from the fact that this element has no square root in P_3. It is interesting that all of the circle groups in P_3 intersect in a pair of points. Thus, in addition to the identity, all circle subgroups in P_3 contain the (central) element which has *minus one* on the main diagonal, *zero's* off the diagonal.

For the moment let Q_3 denote the *universal covering group* of P_3 (this is a group locally isomorphic to P_3 uniquely defined by the condition that it shall be *simply-connected*). There is in Q_3 a discrete central subgroup Z such that Q_3/Z is isomorphic to P_3; actually Z is isomorphic to the group of integers. The circle groups in P_3 are covered by line groups in Q_3 all of which *intersect* in the elements of Z. It was proved by E. Cartan [1] that Q_3 is not isomorphic to a matrix group (of course it is *locally* isomorphic); this is expressed by saying that Q_3 does not admit a *faithful representation* (4.14).

4.13.1. Very much is known about the topological structure of compact groups and their coset-spaces, and more is in constant process of being learned. We shall quote only one theorem (see the article by Samelson, cited above, for references. This theorem is algebraic as well as topological and has a number of applications.

THEOREM. *A compact connected Lie group has a maximal toral subgroup of dimension $r \geq 1$, called the rank r of the group. All such subgroups are conjugate to one another and every element belongs to at least one such subgroup.*

The compact connected three-dimensional Lie groups are: the toral product of three circle-groups, the orthogonal group in three-space (consisting of rotations about the origin), and the universal covering-group of this rotation group. The first of these is abelian. Each of the others is of *rank one*: the maximal toral groups are circle groups (corresponding to *axial* rotations). The covering group has the three-sphere as manifold, the rotation group has the projective three-space; the covering is two-to-one.

4.14. Faithful representations

We shall not consider the subject of faithful representations of

Lie groups (see Goto [2], Matsushima [1], and references there) except to give the following example by Birkhoff [1] of a group not isomorphic to a subgroup of a matrix group. The existence of such groups imposes some difference between the methods available for the analysis of compact groups and of the non-compact ones. The first such example is due to Cartan [1].

Let G_3 be the group described in the preceding section consisting of matrices of the form:

$$M(x, y, z) = \begin{pmatrix} 1 & x & z \\ 0 & 1 & y \\ 0 & 0 & 1 \end{pmatrix}$$

and let N be the invariant subgroup consisting of the matrices $M(0, 0, n)$, n any integer. Then $G_3^* = G_3/N$ is not isomorphic to a matrix group.

The proof is as follows (Birkhoff [1]).

LEMMA. *Let G be any group of linear transformations of an m-dimensional vector space E_m over the complex field. Suppose that G contains elements S and T whose commutator $R = S^{-1}T^{-1}ST$ is of prime order p, and satisfies $SR = RS$, $TR = RT$. Then the dimension of E_m is at least p.*

Since R is of order p, E_m contains a vector x such that $Rx = \alpha x$, where α is a primitive p-th root of unity (see A. Speiser, theorem 127 in 3rd edition, theorem 126 in 2nd. edition). Now let V denote the linear subspace of all vectors x of E_m satisfying $Rx = \alpha x$. If $x \varepsilon V$ then

$$R(Sx) = S(Rx) = S(\alpha x) = \alpha(Sx).$$

This shows that S, and similarly T, transforms V into itself.

Now if we confine our attention to the linear space V, then $S^{-1}T^{-1}ST = \alpha I$ (where I denotes the unit matrix corresponding to the dimension of V). Therefore $T^{-1}ST = \alpha S$. But $T^{-1}ST$ and S have the same characteristic roots; hence so do S and αS as linear transformations of V. Moreover since S is non-singular, its characteristic numbers are not zero; hence it has at least p characteristic numbers. It follows that V and therefore E_m is at least p-dimensional.

This concludes the proof of the Lemma.

The application is as follows. Suppose that G_3^* is isomorphic to an $m \times m$ matrix group. This is a linear group on E_m. Let p be any prime. Let S be the image of M $(1, 0, 0)$ and let T be the image of $M(0, 1/p, 0)$. Both of these commute with $R = S^{-1}T^{-1}ST$ which is the image of $M(0, 0, 1/p)$ in G_3; R is of order p. It follows from the Lemma that $m \geq p$. *Since p is any prime this is impossible.*

4.15. Remarks on the non-metric case

Here we sketch a method for extending the results of 4.8 and 4.9 to the non-separable case. We leave details to the reader. The main tool is Theorem 1.

THEOREM 1. *Let G be a locally compact group and H a compact normal subgroup, and T the natural map from G to G/H. If $x(t)$ is a one-parameter group in G/H there is a one-parameter group $a(t)$ in G such that $Ta(t) = x(t)$.*

For the proof the following Lemmas may be used.

LEMMA 1. *For any G and H as in the theorem,*

$$T(G_0) = (G/H)_0.$$

We omit the proof. This fact shows that in proving Theorem 1 it will be sufficient to consider G connected.

LEMMA 2. *Let G, $x(t)$, H be as in Theorem 1. Let G be connected. Let $K \subset H$ be a compact invariant subgroup of G and let N be a compact subgroup of K such that 1) N is invariant in K 2) K/N has no small groups. Then N is invariant in G. Moreover if G/K contains a one-parameter group mapping into $x(t)$ by the natural map then G/N has the same property.*

The first part of the conclusion depends on the fact that gNg^{-1} is in K and since it must equal N for g small (K/N has no small groups), we always have $g\,Ng^{-1} = N$ because G is connected. For the second part consider the map

$$G/N \to G/K.$$

The kernel is the Lie group K/N, and this makes it possible to prove the last part. We next prove Theorem 1.

We know there is a compact subgroup K (for example $K = H$) satisfying

   a) $K \subset H$; K invariant in G;

   b) G/K contains a one parameter group mapping into $x(t)$ under the natural map of G/K to G/H.

If K is any such group let $y(t)$ be the one-parameter group in G/K which goes to $x(t)$ and let S be the map from G to G/K. Let

$$A(t) = S^{-1}y(t).$$

Then $A(t)$ is a group (not necessarily locally compact) and for each t, $A(t)$ is a coset of K. Furthermore $TA(t) = x(t)$, and $A(R_1)$ is a compact set.

By Lemma 2 any group K of this kind contains a smaller group of this kind if $K \neq e$. We can see also that any transfinite decreasing sequence of such groups has an intersection of the same kind. Therefore there must be a minimal K and this must be e, and for $A(0) = e$, $A(t)$ is a one-parameter group. This proves the Theorem.

4.15.1.   To apply the theorem let G be any locally compact group and H a compact normal subgroup such that G/H is an n-parameter Lie group. In G/H choose n one-parameter groups $x_1(t_1), \ldots, x_n(t_n)$ which can be used for canonical coordinates of the second kind (we do not give the proof that these exist) that is such that

$$x_1(t_1)x_2(t_2) \ldots x_n(t_n), \quad |t_i| \leq r$$

gives an n-cell in a one-one way. Then in G choose $a_1(t_1), \ldots, a_n(t_n)$ such that

$$Ta_i(t_i) = x_i(t_i), \quad i = 1, \ldots, n.$$

Then

$$a_1(t_1)a_2(t_2) \ldots a_n(t_n), \quad |t_i| \leq r$$

must give an n-cell in G in a one-one way, for T is a homomorphism and if the representation above were not one-one then the representation below could not be.

Thus any n-cell in G/H can be lifted to G, and this provides a basis for obtaining the results of 4.8 and 4.9 for the non-separable case. In the abelian case of 4.12 this method enables us to lift a vector group from G/H to G.

For more details on the non-separable case see Iwasawa [1]; this paper was of great importance in several directions in the development of the theory of locally compact groups.

CHAPTER V

Transformation Groups

5.0. Introduction

From now on we shall emphasize properties of transformation groups rather than properties of groups themselves, and summarize a few of the known results.

A transformation group of a space has been defined (1.26) as a pair (G, M) where G is a topological group, M is a space and where further to each element $g \in G$ there is given a homeomorphism $g(x) = f(g; x)$ of M (Hausdorff) onto itself satisfying

1) $f(g; x) = g(x)$ is simultaneously continuous in g and x;

2) $g_1(g_2(x)) = (g_1 g_2)(x)$.

There are several different levels of generality on which questions may be raised and on all of these levels there are many unsolved problems. Most of our attention will be directed to the case where M is a manifold (1.27). In this chapter M will first be a differentiable manifold, and for each fixed g, $f(g; x)$ will be a differentiable homeomorphism. When G is locally compact and effective, it will follow that G is a Lie group. It will be seen also that a considerable number of differentiability properties are implied by rather light assumptions. This fits in with the general observation that group properties often force considerably more regularity than is explicitly postulated.

This chapter also includes a theorem about a general Lie group G which says that if H is a closed subgroup there is a neighborhood O of H such that any closed subgroup in O is conjugate to a subgroup in H (Montgomery-Zippin [8]). This can be regarded as an extension of the theorem that G contains no small subgroup, and it has a number of applications to transformation groups and their geometric properties, some of which are presented in the next chapter.

Later in the chapter there is a section on local cross-sections of orbits and another on periodic and pointwise periodic homeomorphism of manifolds.

Although the first four chapters are rather self contained we shall not hesitate in these last two chapters to make use of standard results from geometry and topology.

5.1. Simultaneous continuity of derivatives

Let M be a manifold with a differentiable (or analytic) structure, and let G be a topological group which acts as a transformation group on M by means of the transforming relation $f(g; x) = g(x)$. If x_0 is a point of M and g_0 is an element of G then for g in some neighborhood of g_0 and x in a neighborhood of x_0 the relation $f(g; x)$ can be expressed in a pair of admissible coordinate systems, one around x_0 and one around $g(x_0)$ by means of a set of n coordinate functions

$$1) \quad f_i(g; x) = f_i(g; x_1, \ldots, x_n), \quad i = 1, \ldots, n.$$

If M is of class C^k (or analytic) and if g is fixed the transformation $g(x)$ is of class C^k (or analytic) provided the functions 1) are of class C^k (or analytic) in admissible coordinate systems in M. We shall consider the case where for each fixed $g \in G$ the transformation $g(x)$ is of class C^k or analytic.

From the fact that $f(g; x)$ is continuous simultaneously in (g, x) and that for each fixed g the partial derivatives are continuous functions of x it is not immediately clear that these partial derivatives are simultaneously continuous in $(g; x)$, but this will be shown to be true. The partial derivatives of $f_i(g; x)$ will be denoted by $f_{ij}(g; x)$, where this expression depends on and is defined only in a pair of admissible coordinate systems covering neighborhoods in M. Of course if the functions $f_{ij}(g; x)$ are continuous in admissible coordinate systems for given neighborhoods they will also be continuous in any other admissible coordinate systems for the neighborhoods. Thus the continuity of $f_{ij}(g; x)$ is independent of which admissible coordinates are chosen.

THEOREM 1. *Let G be a locally compact group which is a transformation group of a manifold M of class C^1. If for each $g \in G$ the transformation $g(x) = f(g; x)$ is of class C^1 then the derivatives of the local coordinate functions $f_i(g; x)$ (these derivatives are $f_{ij}(g; x)$) are continuous in $(g; x)$ simultaneously.*

In the proof (Montgomery [2]) use will be made of methods of Baire and we begin by proving one of his results formulated so as to apply here. Theorem 1 is true also when G is a complete metric space or more generally whenever category arguments (1.7.3) are valid but we do not formulate or prove this extension of the theorem.

LEMMA 1 (BAIRE). *Let $\varphi_n(g)$ be a convergent sequence of continuous real functions on a compact Hausdorff space Z with $\varphi(g) = \lim \varphi_n(g)$. Then points of continuity of $\varphi(g)$ are everywhere dense.*

Let E_i be the set of points where $\varphi(g)$ has a discontinuity $\geq 1/i$, that is g is in E_i if and only if for every positive ϵ, every neighborhood of g contains two points on which φ differs by at least $1/i - \epsilon$. The set E_i is a closed subset of Z. The union of the sets E_i is the set of discontinuities of φ and it is sufficient to show that E_i includes no interior points, for then $\cup E_i$ is of the first category and hence $Z - \cup E_i$ is everywhere dense.

Assume that some E_i contains a non-null open set E and let E_{iN} for each $N \in I$ be the points g of E_i such that $|\varphi_n(g) - \varphi_m(g)| \leq 1/3i$, $m, n \geq N$. For g in E_{iN},

$$|\varphi_n(g) - \varphi(g)| \leq 1/3i, \quad n \geq N.$$

The set E_{iN} is a compact subset of E_i and

$$\cup_N E_{iN} = E_i.$$

Then for some one integer N, E_{iN} contains a point a in E which is an inner point of E_{iN} by a category argument. Hence there is a neighborhood A of a such that if b and c are in A then these points are in E_{iN}. Therefore for any b and c in A,

2) $|\varphi(b) - \varphi_n(b)| \leqq 1/3i, \quad n \geqq N$

3) $|\varphi_n(c) - \varphi(c)| \leqq 1/3i, \quad n \geqq N.$

However by the definition of E_i we must have for some b and c in A and for a preassigned positive ϵ

4) $|\varphi(c) - \varphi(b)| \geqq 1/i - \epsilon.$

These inequalities imply

$$|\varphi_n(c) - \varphi_n(b)| \geqq 1/3i - \epsilon$$

but b and c may be chosen in any neighborhood of a, that is φ_n is discontinuous at a, contrary to hypothesis. Hence the assumption that E_i contains an inner point leads to a contradiction. Therefore E_i contains no interior point and the lemma is proved.

LEMMA 2. *Let Z be compact Hausdorff and let V be the set $x_1^2 + \cdots + x_n^2 < 1$ in euclidean n-space. Let $F(g; x_1, \ldots, x_n)$ be a continuous real valued function defined on $Z \times V$ and let $F_j(g; x) = (\partial/\partial x_j)F(g; x_1, \ldots, x_n)$ exist and be continuous in x for any fixed g in Z. If a is a point of V then the points g_0 such that $F_j(g; x)$ is simultaneously continuous at (g_0, a) are everywhere dense in Z (j fixed). The set where this is false is first category.*

It must be proved that there is one of the required points in any compact neighborhood Z^* of Z. However the argument for Z^* instead of for Z involves only a change of notation. This will not be made since we now see in this way that it will be sufficient to show that Z must contain at least one such point.

For a given positive r and each g in Z choose $S(a, g, r)$ to be an open set in V which includes a and is such that if x is in $S(a, g, r)$ then

$$|F_j(g; x) - F_j(g; a)| \leqq r/4.$$

The set $S(a, g, r)$ must exist because $F_i(g; x)$ is continuous in x for each fixed g by hypothesis. Let O_n be a sequence of neighborhoods of a in V which have a as their only common point. Let G_n be the collection of points g of Z such that

$$O_n \subset S(a, g, r).$$

It follows that

$$\cup_n G_n = Z.$$

Hence by a category argument there is a value of n, say n_1, such that G_{n_1} is everywhere dense in an open subset Z_1 of Z, that is $(G_{n_1} \cap Z_1)$ is dense in Z_1.

Then if x is in O_{n_1} and g is G_{n_1},

5) $| F_j(g; x) - F_j(g; a) | \leq r/4.$

For a fixed, $F_j(g; a)$ is a limit of a sequence of continuous functions of g and by the preceding Lemma there must be a point g_0 in Z_1 where $F_j(g; a)$ is continuous in g. There is an open set Z_1^*,

$$Z_1^* \subset Z_1 \subset Z,$$

such that if g is in Z_1^* then

6) $| F_j(g; a) - F_j(g_0; a) | \leq r/4.$

By the two inequalitites above it can be seen that for $x \in O_{n_1}$ and $g \in G_{n_1} \cap Z_1^*$

7) $| F_j(g; x) - F_j(g_0; a) | \leq r/2.$

Consider the quantity

8) $(1/\Delta x_j)[F(g; x_1, \ldots, x_j + \Delta x_j, \ldots, x_n) - F(g; x_1, \ldots, x_n)].$

By the theorem of the mean this quantity is equal to the following for some t, $0 < t < 1$,

$$F_j(g; x_1, \ldots, x_j + t\Delta x_j, \ldots, x_n).$$

Hence if

i) $x \in O_{n_1}$
ii) $g \in G_{n_1} \cap Z_1^*$
iii) Δx_j is so small that the closed interval from $(x_1, \ldots, x_n)$ to $(x_1, \ldots, x_j + \Delta x_j, \ldots, x_n)$ is in O_{n_1}, then

9) $| (1/\Delta x_j)[F(g; x_1, \ldots, x_j + \Delta x_j, \ldots, x_n) - F(g; x)]$
     $- F_j(g_0; a) | \leq r/2.$

The difference quotient is a continuous function of all variables entering into it and since the set $(G_{n_1} \cap Z_1^*) \times O_{n_1}$ is dense in $Z_1^* \times O_{n_1}$ it follows that the relation above holds for x in O_{n_1} and g in Z_1^*.

The difference quotient in 9) has $F_j(g; x)$ as a limit and consequently 7) also holds for $x \epsilon O_{n_1}$, $g \epsilon Z_1^*$.

We shall now summarize what has been done so far to prove the Lemma. It has been shown that given $a \epsilon V$ there is an open set $O_{n_1} \subset V$ including a and an open set Z_1^* in Z such that if (g, x) and (g', x') are in $Z_1^* \times O_{n_1}$ then

$$9') \quad | F_j(g; x) - F_j(g'; x') | \leqq r.$$

By a similar procedure there can be found an O_{n_2} with

$$a \epsilon O_{n_2} \subset O_{n_1} \subset V$$

and an open set Z_2^*, whose closure is in Z_1^*

$$\overline{Z}_2^* \subset Z_1^*,$$

giving an inequality similar to the above 9') but with r replaced by $r/2$. Continuing in this way a point of the kind desired may be found by choosing any point in $\cap Z_i^* = \cap \overline{Z}_i^*$. This completes the proof except for the last statement.

The points of continuity of a function are a G_δ (= intersection of a countable collection of open sets) and if such a set is everywhere dense then the complement is of the first category (= union of a countable collection of nowhere dense sets). Therefore the points g which do not satisfy the requirement are a set of the first category. This completes the proof.

In the application to be made of the above Lemma there are n functions $f_i(g; x)$ and they have partial derivatives $f_{ij}(g; x)$ which are continuous in x for g fixed. The Jacobian of the transformation $f(g; x)$ is the determinant of the matrix $(f_{ij}(g; x))$

$$\det \, (f_{ij}(g; x))$$

and is denoted by $D(g; x)$; the functions f_i, f_{ij}, and $D(g; x)$ of course depend on the admissible coordinate systems chosen, but the continuity of all these functions is independent of which admissible systems are chosen.

Let a in M be given and let B_{ij} be the set of points g_0 in G such that $f_{ij}(g; x)$ has a discontinuity at (g_0, a). The union on i and j of B_{ij} is a set of the first category. This proves the following:

LEMMA 3. *Under the hypothesis of Theorem* 1 *above let a be a point of M. Then the set of points* g_0 *in G such that all the functions* $f_{ij}(g; x)$ *and hence also* $D(g; x)$ *are continuous at* $(g_0; a)$ *is everywhere dense in G.*

There is one more Lemma which will also be useful in the proof of Theorem 1.

LEMMA 4. *Under the hypothesis of Theorem* 1 *let a be a point of M. Then for all i, j the function* $f_{ij}(g; x)$ *is simultaneously continuous at* $(e; a)$ *where e is the identity of G.*

Let g_0 be an element of G such that the functions $f_{ij}(g; x)$ for all i, j are continuous at $(g_0; a)$. The function $D(g; x)$ is also continuous at $(g_0; a)$; since each transformation in G has an inverse, $D(g; x)$ can never be zero; in particular it cannot be zero at $(g_0; a)$ and therefore $D(g; x)$ must be bounded away from zero in some neighborhood of $(g_0; a)$.

If h is in a sufficiently small neighborhood of e then g_0h is in a preassigned neighborhood of g_0. For such h and for x near a the following relations have meaning and are true:

$$f_i(g_0h; x_1, \ldots, x_n) = f_i[g_0; f_1(h; x), \ldots, f_n(h; x)].$$

Hence

10) $f_{ij}(g_0h; x) = \Sigma_k f_{ik}[g_0; f(h; x)]f_{kj}(h; x).$

When $h = e$ and $x = a$ this becomes

11) $f_{ij}(g_0; a) = \Sigma_k f_{ik}(g_0; a)\delta_{kj}.$

When h is near e and x is near a then $f_{ij}(g_0h; x)$ is near $f_{ij}(g_0; a)$ and $f_{ik}[g_0; f(h; x)]$ is near $f_{ik}(g_0; a)$. The first of these things is true because of the simultaneous continuity at $(g_0; a)$; the second is true because of continuity in the right hand variable at an arbitrary point. Furthermore the determinant whose elements are $f_{ik}[g_0; f(h; x)]$ is bounded away from zero for x near a and h near e.

With j held fixed, 10) may be solved for $f_{kj}(h; x)$ and the above remarks show for x near a, h near e, that $f_{kj}(h; x)$ is near δ_{kj} which is equal to $f_{kj}(e; a)$. This concludes the proof of the Lemma.

The proof of Theorem 1 will now be given. Let a be any point of M and let g be any element of G. By a property of a transformation group

$$f(gh; x) = f[g; f(h; x)],$$

and in suitably chosen neighborhoods of g, of e (h near e), of a (x near a)

$$f_i(gh; x) = f_i[g; f(h; x)].$$

From this

$$f_{ij}(gh; x) = \Sigma_k f_{ik}[g; f(h; x)]f_{kj}(h; x).$$

When h is near e and x is near a, $f_{kj}(h; x)$ is near $f_{kj}(e; a) = \delta_{kj}$ by the preceding Lemma; also under the same conditions $f_{ik}[g; f(h; x)]$ is near $f_{ik}[g; a]$ by the continuity in the right hand variable. From this it can be seen that $f_{ij}(gh; x)$ is near $\Sigma_k f_{ik}[g; a]\delta_{kj} = f_{ij}(g; a)$, but this asserts that the function is simultaneously continuous at $(g; a)$ and this completes the proof of the Theorem.

COROLLARY. *If G is a locally compact group which is a transformation group of a manifold M of class C^k and if each transformation of the group is of class C^k then the derivatives of the local functions $f_i(g; x)$ of order up to and including k are simultaneously continuous in $(g; x)$.*

The proof will be made by induction so it will be assumed that the theorem is true for $k - 1$. Then by this assumption all partial derivatives of order up to and including $k - 1$ of the functions $f_i(g; x)$ are simultaneously continuous in $(g; x)$.

Let a be any point of M. The set of elements g_0 in G such that a given kth order derivative is simultaneously continuous at $(g_0; a)$ is an everywhere dense set whose complement is of the first category. This follows from Lemma 2. Since a finite number of functions is involved the same remark holds for the g_0's such that *all* the kth order partial derivative are continuous at $(g_0; a)$.

The proof may now be made by following along the lines of the proof of Lemma 4 and the subsequent proof of Theorem 1. Carrying out the details for $k = 2$ will suggest to the reader the general procedure to be followed.

5.2. Differentiability with respect to group parameters and local linearity at fixed points

The following Lemma was proved by H. Cartan [1].

LEMMA 1. *Let G be a topological group which acts on a manifold M of class C^1 and assume for each g that $g(x) = f(g; x)$ is of class C^1 in x. Let W_1 and W_2 be open admissible convex coordinate neighborhoods in M, $\overline{W}_1 \subset W_2$, and let U be a neighborhood of e in G such that $\overline{U}\,\overline{W}_1 \subset W_2$ and let g, g^2, ..., g^q be in $\overline{U}$. Then for all x in $\overline{W}_1$,*

$$f_i(g^q; x) - x_i = \Sigma_j[\delta_{ij}(g; x; q)]\,[q\{f_j(g; x) - x_j\}],$$

where with $y_i(g; x) = f_i(g; x) - x_i$,

$$\delta_{ij}(g; x; q) = (1/q) \int_0^1 [\delta_{ij} + f_{ij}(g; x + ty) + \ldots + f_{ij}(g^{q-1}; x + ty)]dt.$$

For the proof let

$$T_i(g; x) = x_i + f_i(g; x) + \ldots + f_i(g^{q-1}; x)$$

and as usual let T_{ij} denote the partial of T_i with respect to x_j. Then (using x for the coordinates of x, etc.)

1) $T_i(g; x + y) - T_i(g; x) = \Sigma_j y_j \int_0^1 T_{ij}(g; x + ty)dt.$

From the definition of T_i we see that

$$T_i(g; x + y) = T_i(g; f(g; x))$$
$$= f_i(g; x) + \ldots + f_i(g^q; x)$$

and therefore

2) $T_i(g; x + y) - T_i(g; x) = f_i(g^q; x) - x_i.$

We now combine 1) and 2) and obtain

$$f_i(g^q;\, x) - x_i = \Sigma_j\,[\textstyle\int_0^1\{\delta_{ij} + f_{ij}(g;\, x + ty) + \ldots$$
$$+ f_{ij}(g^{q-1};\, x + ty)\}dt][f_j(g;\, x) - x_j].$$

Multiplying and dividing on the right by q gives the desired result and thus completes the proof of the Lemma.

REMARK.   By using the above Lemma we can prove that if $a(t)$ and $b(t)$ are one-parameter groups in a Lie group G, then

$$\lim\,(a(t/q)\,b(t/q))^q$$

exists at least for t in a neighborhood of zero. (Compare 2.16 and note that this proves that in Chapter III the convergence is in actuality for the full sequence). In this use of Lemma 1, $f(g;\, x)$ is to represent group multiplication in a Lie group G. It should be noted that $f_{ij}(g;\, x)$ is then uniformly near δ_{ij} for g ranging over any compact set and x in a sufficiently small neighborhood of e; we take e to have coordinates zero. The element g has coordinates, and in our application of Lemma 1, $x = 0$, and y becomes g. The groups $a(t)$ and $b(t)$ are given in coordinates by

$$a_i(t) = a_i t + c_i t^2 + \ldots$$
$$b_i(t) = b_i t + d_i t^2 + \ldots$$

and hence

$$(a(t)b(t))_i = (a_i + b_i)t + A_i(t)t$$

where $A_i(t)$ approaches zero with t. Letting $g = a(t/q)b(t/q)$ in the Lemma gives

$$f_i((a(t/q)b(t/q))^q;\, 0) = (a(t/q)b(t/q))_i^q =$$
$$= \Sigma_j[\delta_{ij}(g;\, 0;\, q)][(a_j + b_j)t + A_j(t/q)t].$$

An examination of the expression for $\delta_{ij}(g;\, 0;\, q)$ shows that it tends to δ_{ij} when q tends to infinity. Hence the above expression has a limit as q tends to infinity and this limit is $(a_i + b_i)t$.

We next deduce two corollaries and later show that any locally compact group of differentiable transformations of a manifold is a Lie group. It is an open question whether or not this is true without differentiability. Even with differentiability the behavior of a compact transformation group in the large presents many difficult topological questions.

COROLLARY 1. *Under the hypothesis of Lemma* 1 *and with the further assumption that the functions* $f_{ij}(g; x)$ *are continuous in* $(g; x)$ *it follows that for every* $\epsilon > 0$, $\overline{U}$ *may be so chosen that if* $g, \ldots, g^q \epsilon \overline{U}$, *then*

$$q[f_j(g; x) - x_j] = \Sigma_i \alpha_{ji}(g; x; q)[f_i(g^q; x) - x_i],$$

where $\alpha_{ji}(g; x; q)$ *is inverse to* $\delta_{ij}(g; x; q)$ *and* $|\alpha_{ji}(g; x; q) - \delta_{ji}| < \epsilon$.

Note that the hypothesis of simultaneous continuity of the function $f_{ij}(g; x)$ is automatically true if G is locally compact by Theorem 1 of the last section.

If U is sufficiently small the simultaneous continuity implies that $\delta_{ij}(g; x; q)$ is near the Kronecker δ_{ij} so that the inverse exists and is also near δ_{ij}. This is true independently of q.

COROLLARY 2. *Under the hypothesis of Corollary* 1 *it follows that if all powers of g are in* $\overline{U}$ *then for any x in* $\overline{W}_1$

$$g(x) = f(g; x) = x.$$

This will be used to prove in Theorem 2 that if G as above is locally compact and effective then G cannot have small subgroups. As a tool for later use we now prove the following:

LEMMA 2. *If G is locally compact and M is (real) analytic and if each transformation of G is analytic then the power series expansions of f(g; x) for g in some neighborhood of e and x in some admissible coordinate system on M are dominated in sufficiently small neighborhoods by convergent power series which are independent of g.*

Let U be a neighborhood of e in G and (x) a fixed coordinate system in M. For each $g \epsilon U$, $f_i(g; x)$ is represented around the origin of (x) by a convergent expansion as follows:

a) $f_i(g; x) = \Sigma a^i_{\lambda_1 \ldots \lambda_n}(g)x_1^{\lambda_1} \ldots x_n^{\lambda_n}$, and for each $k = 1, 2, \ldots$ we denote by E_k the subset of U for which

b) $\sup_{i, \lambda_1, \ldots, \lambda_n} \left| a^i_{\lambda_1 \ldots \lambda_n}(g) / k^{\lambda_1 + \cdots + \lambda_n} \right| \leq k$.

Since for each g, a) has some domain of absolute convergence, U is the union of all E_k. Also the coefficients $a_\lambda(g)$ are values at $x = 0$ of partial derivatives of $f(g; x)$ with respect to x, and thus by an earlier corollary are continuous. Therefore each E_k is closed in U and by a category argument, at least one E_k contains an inner point, call it g_0. Since, for fixed k, the inequalities b) imply dominated convergence of the series a) in some x-neighborhood, our theorem has been proved for some sub-neighborhood $U(g_0)$ of U.

The group relations

$$f_i(g_0^{-1} g; x) = f_i(g_0^{-1}; f(g; x))$$

will permit these majorants to be transferred to some neighborhood of e, or any other element. This completes the proof of the Lemma.

The above lemma sometimes makes it possible to integrate $f_i(g; x)$ over G and be sure the result is analytic. For suppose G is compact and leaves fixed the origin of a coordinate system x in a manifold M. Further assume for each g that $f(g; x)$ is analytic. The Lemma above then shows that the expansion for $f_i(g; x)$ around the origin are majorized by a convergent series. Hence integrating over G, that is, forming

$$\int_G f_i(g; x) dg$$

leads to an analytic function of x, defined by a convergent power series in a neighborhood of the origin of (x).

This will be used later and in fact a closely related application is needed in the following proof.

The following theorem was proved by Bochner [1]. A *stationary point* is one left fixed by every element of the group.

THEOREM 1. *Let G be a compact group of transformations of a manifold M of class C^k ($k \geq 1$) or analytic and let each transfor-*

*mation of G be of class C^k or analytic. Then in the vicinity of a
stationary point admissible coordinates may be so chosen that the
transformations are linear.*

Because G is compact the stationary point is in arbitrarily
small neighborhoods which are invariant under G. Inside one of
these we choose a *convex* admissible coordinate system with
origin at the stationary point. The theorem of the mean shows
that if $T(g)$ denotes $f(g; x)$ then

$$T_i(g) = f_i(g; x) = \sum_{j=1}^{n} a_{ij}(g)x_j + \sum_{j=1}^{n} b_{ij}(g; x)x_j$$

where $b_{ij}(g; x)$ tends to zero with $\sum x_i^2$.

Let $L(g) = (a_{ij}(g))$ be the linear transformation determined by
the first term above and we then have

$$L(g)L(h) = L(gh).$$

For each $g \in G$ consider the transformation $L(g^{-1})T(g)$ which is
given by the function $h(g; x)$ where

$$h_i(g; x) = \sum_{j=1}^{n} a_{ij}(g^{-1})f_j(g; x).$$

We see also that

$$h_i(g; x) = x_i + \sum_j c_{ij}(g; x)x_j$$

where $c_{ij}(g; x)$ tends to zero with $\sum x_i^2$.

We consider now the transformation R defined as follows:

$$R = \int_G L(g^{-1})T(g)dg$$

where the integration is normalized so that the volume of the
whole group is one. Then

$$R_i(x_1, \ldots, x_n) = \int_G h_i(g; x)dg;$$

furthermore R is of class C^k and it has an inverse of class C^k
in a neighborhood of the origin (in the analytic case R and its
inverse are analytic). Let a be an element of G so that

$$L(a)R = L(a) \int L(g^{-1})T(g)dg$$
$$= \int L(ag^{-1})T(g)dg.$$

Changing variables by letting $ag^{-1} = h^{-1}$, $g = ha$ gives

$$L(a)R = \int L(h^{-1})T(ha)dh$$
$$= \int L(h^{-1})T(h)dh \cdot T(a) = RT(a).$$

Therefore

$$L(a) = RT(a)R^{-1}$$

that is R, which is an admissible transformation, transforms $T(g)$ into $L(g)$ which is linear. This completes the proof of the Theorem.

THEOREM 2. *Let G be a locally compact effective transformation group of a connected manifold M of class C^1 and let each transformation of G be of class C^1. Then G does not contain small subgroups and by earlier results G must be a Lie group.*

By Corollary 2 above any sufficiently small subgroup must leave some maximal open set Q stationary; $\overline{Q}$ is also stationary. By Theorem 1, G is locally linear around a stationary point and therefore this stationary set Q has no boundary and hence we can see that it includes all of space. Thus any sufficiently small group leaves all of M fixed. Since G is effective such a small group must contain only the identity, that is there are no small groups. This completes the proof.

The above fact was proved in Bochner-Montgomery [2] in the C^2 case, and proved by Kuranishi [2] in the C^1 case. Earlier Myers and Steenrod [1] proved a related theorem for isometries of Riemannian manifolds and H. Cartan for bounded domains in complex variables. We show (Bochner-Montgomery [1]) that the transforming functions automatically have derivatives with respect to the group parameters; this now has meaning since G is a Lie group. It will be shown in the analytic case that the transforming functions automatically are analytic in the group parameters (loc. cit.).

THEOREM 3. *Let G be a Lie group which acts as a transformation group on a manifold M of class C^1 in such a way that for each a in G, $f(a; x)$ is of class C^1. If $a = (a_1, \ldots, a_r)$ are analytic para-*

meters for G then $(\partial/\partial a_j)f_i(a; x)$ *exists and is simultaneously continuous in* $(a; x)$, $i, j = 1, \ldots, n$.

5.2.1. The proof of Theorem 3 will be made in several steps which we shall state as Lemmas (Bochner-Montgomery [1]).

LEMMA A. *If the G of Theorem 3 is a one-parameter group with s as parameter and* $f_i(s; x)$ *as the transforming functions then the derivative* $(\partial/\partial s)f_i(0; x)$ *exists and for small h*

a) $f_i(h; x) - f_i(0; x) = \Sigma_j [\int_0^h f_{ij}(t; x)dt][(\partial/\partial s)f_j(0; x)]$.

The following function will be used in the proof:

b) $T_i(h; x) = \int_0^h f_i(t; x)dt$.

Because f_{ij} has been shown to be simultaneously continuous it follows that differentiation under the integral sign is permissible in $b)$, that is $T_{ij} = (\partial/\partial x_j)T_i$ exists and

c) $T_{ij}(h; x) = \int_0^h f_{ij}(t; x)dt$.

Using the mean value theorem gives

d) $T_i(h; y) - T_i(h; x) = \Sigma_j T_{ij}(h; \bar{x}(i)) (y_j - x_j)$

where $\bar{x}(i) = x + \theta_i(y - x)$, $0 < \theta_i < 1$.
Let p be a small real number and let $y = f(p; x)$. Then

e) $T_i(h; y) - T_i(h; x) = \int_0^h f_i(p + t; x)dt - \int_0^h f_i(t; x)dt$
$= \int_p^{h+p} f_i(t; x)dt - \int_0^h f_i(t; x)dt$
$= \int_h^{h+p} f_i(t; x)dt - \int_0^p f_i(t; x)dt$.

Substituting c) and d) in e) and dividing by p gives

f) $(1/p) \int_0^p [f_i(h + t; x) - f_i(t; x)]dt$
$= \Sigma_j [\int_0^h f_{ij}(t; \bar{x}(i)) dt] [(1/p)(y_j - x_j)]$.

Since $f_{ij}(t; x)$ is simultaneously continuous and equal to δ_{ij} when $t = 0$, the matrix

$$\| (1/h) \int_0^h f_{ij}(t; \bar{x}(i)) dt \|$$

is near the identity matrix for h small and this is true uniformly for x in a compact neighborhood and h in a closed interval. For this reason f) can be solved for $(1/p)(y_j - x_j)$. Let $\| \varphi_{ij}(h; \bar{x}(i)) \|$ be the inverse of the matrix

$$\| \textstyle\int_0^h f_{ij}(t; \bar{x}(i))\, dt \|.$$

If $y_j - x_j$ is replaced by its value we obtain as the solution of f) the following

g) $(1/p)[f_j(p; x) - x_j] =$

$$\Sigma_i \varphi_{ij}(h; \bar{x}(i))\, (1/p) \textstyle\int_0^p [f_i(t + h; x) - f_i(t; x)]dt.$$

The right hand side of g) has a limit as p tends to zero. But the limit of the left hand side is $(\partial/\partial s)f_j(0; x)$ so that the existence of this quantity is established.

LEMMA B. *Let the G of Theorem 3 have r parameters $g = (a_1, \ldots, a_r)$ where these are canonical coordinates such that the curves*

$$\gamma_1 = b_1 t, \ldots, \gamma_r = b_r t$$
$$b_1^2 + \ldots + b_r^2 = 1$$

are local one-parameter groups. If we introduce the functions

1) $f_i(t; b; x) = f_i(bt; x)$

pertaining to such a subgroup then the derivatives

2) $F_i(b; x) = (\partial/\partial t)f_i(0; b; x)$

are continuous in $(b_1, \ldots, b_r; x_1, \ldots, x_n)$.

The formula a) of Lemma A holds for any one-parameter group, so that in a), $f_i(t; x)$ may be replaced by $f_i(t; b; x)$ and we may then invert to obtain

3) $F_j(b; x) = \Sigma_i \varphi_{ij}(h; b; x)[f_i(h; b; x) - f_i(0; b; x)]$

where the matrix $\| \varphi_{ij} \|$ is the inverse of

4) $\| \textstyle\int_0^h f_{ij}(t; b; x)dt \|.$

However for a fixed small h, 3) is continuous in $(b; x)$ so that 2) is also, and this proves the Lemma.

LEMMA C. *For the parameters* $g = (a_1, \ldots, a_r)$ *chosen as in Lemma B the partial derivatives*

$$(\partial/\partial a_\varrho) f_i(a_1, \ldots, a_r; x_1, \ldots, x_n), \quad \varrho = 1, \ldots, r$$

exist and are continuous in $(a; x)$.

The proof of this Lemma which will now be given will complete the proof of Theorem 3. It will be sufficient to consider only the parameter a_1. We consider the group elements

$$\begin{aligned} g &= (a_1, \ldots, a_r) \\ g_h &= (a_1 + h, \ldots, a_r) \\ \gamma &= g_h g^{-1} = \gamma(a; h). \end{aligned}$$

Let $\gamma = (\gamma_1, \gamma_2, \ldots, \gamma_r)$ and then these components of γ have the following form:

$$\begin{aligned} \gamma_1 &= h + Q_2(h; a) \\ \gamma_\sigma &= Q_2(h; a), \quad \sigma = 2, \ldots, r \end{aligned}$$

where $Q_2(h; a)$ is a symbol representing the appropriate power series in $(h; a_1, \ldots, a_r)$; these series are divisible by h and contain only terms of degree two or more. It can be seen that we can write

$$\gamma_\varrho = b_\varrho (h; a) \cdot t(h; a), \quad \varrho = 1, \ldots, r,$$

with

$$\Sigma b_\varrho^2 = 1,$$

where $b_\varrho(h; a)$ and $t(h; a)$ are power series in $(h; a)$ and $b_1(h; a)$ is normalized to have the form

$$1 + Q_1(h; a)$$

with $Q_1(h; a)$ being a power series without a constant term.

Since $f(g_h; x) = f(\gamma; g(x)) = f(\gamma; f(g; x))$, the difference quotient

5) $\quad (1/h)[f_i(a_h; x) - f_i(a; x)]$

has the value

$$(1/h)[f_i(\gamma; a(x)) - f_i(0; a(x))]$$

and by a) of Lemma A this is equal to

6) $\quad \Sigma_j [(1/h) \int_0^{t(h;a)} f_{ij}(\tau; b(h; a); a(x)) d\tau] F_j(b(h; a); a(x))$

where F_j is defined by 2). Now

$$f_{ij}(\tau; b(h; a); a(x)) = f_{ij}(\tau; b(h; a); f(a; x))$$

is continuous in $(\tau; h; a; x)$ and $F_j(b(h; a); a(x))$ is continuous in $(h; a; x)$ by Lemma B. Therefore we may let h tend to zero in 6) and the limit is

$$[(\partial/\partial h)t(0; a)] \cdot \Sigma_j f_{ij}(0; b(0; a); a(x)) F_j(b(0; a); a(x)).$$

This however is continuous in $(a; x)$.

LEMMA. *Let G be a Lie group acting on a manifold M of class C^k (analytic) and let each transformation of G be of class C^k (analytic). If $a = (a_1, \ldots, a_r)$ are analytic parameters for G then $(\partial/\partial a_j)f_i(0; x)$ are class C^{k-1} (analytic) functions of x.*

It will again be convenient and involve no loss of generality to consider the index $j = 1$. In a) of Lemma A above let $a = 0$ and obtain

7) $\quad f_i(h, 0, \ldots, 0; x) - f_i(0, 0, \ldots, 0; x) =$
$$\Sigma_j [\int_0^h f_{ij}(t, 0, \ldots, 0; x)dt] \cdot [(\partial/\partial a_1)f_j(0; x)].$$

Consider first the analytic case. The functions $f_i(h, 0, \ldots, 0; x)$ are analytic in x for a fixed h and $f_i(0; x)$ is also analytic. Furthermore

$$\int_0^h f_{ij}(t, 0, \ldots, 0; x)dt$$

is analytic in x because by an earlier result it follows that the power series entering under the integral sign are majorized. As before the equations above can be solved for $(\partial/\partial a_1)f_j(0; x)$ and it follows that this quantity is analytic.

Consider next the case where $f(a; x)$ is of class C^k in x for each fixed a. In this case the same equations (7) are used. They may be solved to express $(\partial/\partial a_1)f_j(0; x)$ in terms of $f_i(h, 0, \ldots, 0, x)$ and $f_i(0; x)$ which are of class C^k and in terms of

$$\int_0^h f_{ij}(t, 0, \ldots, 0; x)dt$$

which is of class C^{k-1}. This proves the lemma.

THEOREM. *Let G be a Lie group which acts as a transformation* *group on a manifold of class C^k (analytic) and assume that for each g, f(g; x) is of class C^k (analytic) in x. If $a = (a_1, \ldots, a_r)$ are analytic parameters for G then the functions defining f(a; x) are of class C^k (analytic) in (a; x) simultaneously.*

It is sufficient to verify the theorem in any neighborhood of the identity in G as it then follows for all of G by group translation. The parameters for G are assumed canonical of the kind used in the proofs above.

For $k = 1$ the theorem has already been proved above. Making use of this will make it possible to derive Lie's differential equations (see Eisenhart [1]) by the usual procedure and this will now be carried out.

Consider the group elements a, b, c subject to the relation

$$ba = c.$$

Let $x' = f(a; x)$ which gives the relation

$$f_i(b; x') = f_i(c; x)$$

in which x, a, and c will be taken as independent variables. Differentiating this relation with respect to a_ϱ gives

8) $$\sum_\sigma \frac{\partial f_i(b; x')}{\partial b_\sigma} \frac{\partial b_\sigma}{\partial a_\varrho} + \sum_j \frac{\partial f_i(b; x')}{\partial x'_j} \frac{\partial x'_j}{\partial a_\varrho} = 0$$

where the rules of differentiation are justified by the results above.

Since $\partial f_i/\partial x'_j$ is continuous in $(g; x)$ and since for $g = e$

$$\frac{\partial f_i}{\partial x'_j} = \frac{\partial f_i(e; x')}{\partial x'_j} = \delta_{ij}$$

the equations 8) can be solved for $\partial x'_j/\partial a_\varrho$ and the solution gives

$$\frac{\partial x'_i}{\partial a_\varrho} = \sum_\sigma \xi_{i\sigma}(b; x') \, \Omega_{\sigma\varrho}(a; b).$$

Letting $b = 0$,

$$\frac{\partial x'_i}{\partial a_\varrho} = \sum_\sigma \xi_{i\sigma}(0; x') \, \Omega_{\sigma\varrho}(a; 0)$$

and these are classical equations:

9) $$\frac{\partial f_i(a;\ x)}{\partial a_\varrho} = \Sigma_\sigma \xi_{i\sigma} \left(f(a;\ x)\right) \Omega_{\sigma\varrho}(a).$$

In these equations the functions $\Omega_{\sigma\varrho}(a)$ are analytic in a since they involve the operations within G itself. The functions $\xi_{i\sigma}(x')$ are analytic in x' if for each fixed g, $f(g;\ x)$ is analytic, by the results above. The functions $\xi_{i\sigma}(x')$ are of class C^{k-1} in x' if $f(g;\ x)$ is of class C^k in x.

If one makes the inductive assumption that $f(a;\ x)$ is of class C^{k-1} in $(a;\ x)$ then it is possible to take $(k-1)$ successive mixed derivatives of the right side of 9) and therefore $f(a;\ x)$ is of class C^k in $(a;\ x)$. This completes the proof for all k for the C^k case.

Next assume that the functions

10) $x_i' = f_i(a;\ x)$

are analytic in x so that the functions $\xi_{i\sigma}(x')$ are analytic in x'. The functions Ω are analytic in a and therefore the functions 10) satisfy the analytic differential equations

11) $$\frac{\partial x_i'}{\partial a_\varrho} = \Sigma_\sigma \xi_{i\sigma}(x') \Omega_{\sigma\varrho}(a).$$

As has been shown the functions 10) are at any rate of class C^k in $(a;\ x)$ for every k. They are also a *complete* solution. But if the system has such a complete solution, then the system is *integrable* and the complete solution is analytic in $(a;\ x)$. This proves the result in the analytic case and completes the proof of the theorem.

In the case of real parameters and complex space coordinates there are analogous results (Bochner-Montgomery [1]).

5.2.2.   Consider now the case where M is a compact complex manifold.   For the remarks to follow see Bochner and Montgomery [3]; we omit proofs. Let G be the collection of all complex analytic homeomorphisms of M onto itself. It can be shown that for each $g \in G$, g^{-1} is also analytic, and thus that G is a group.

The group G can be topologized by giving M a metric and then giving G the metric

$$D(g_1, g_2) = \max d(g_1(x), g_2(x)).$$

With this metric it can be shown that G is locally compact; by a theorem above it follows that G is a real Lie group with parameters (a) and that the transforming functions $f(a; x)$ are automatically differentiable with respect to (a). It can further be shown that G is a complex Lie group.

In case M is a bounded domain in the space of several complex variables, let G be the group of all complex analytic homeomorphisms of M onto itself. It follows in a similar way that G is a Lie group. For this case the theorem was first proved by H. Cartan [1] who introduced some of the devices used and extended in Bochner-Montgomery [1].

5.3. Conjugacy of subgroups

The theorem to be proved next is of interest in itself and has applications to the action of Lie groups as transformation groups. When a compact group G acts as a transformation group of a space M and if $x \in M$ and if O is any open set in G including G_x (G_x = elements g such that $g(x) = x$) then for y sufficiently near to x, G_y is in O. For this reason the relation between one closed subgroup of G and another in an open neighborhood of the first is of help in understanding the action of groups.

Let F be a compact subgroup of any Lie group G, and let M be the homogeneous space G/F,

$$M = G/F.$$

The group F acts on G/F and leaves $x = F$ fixed; it acts on the tangent space of G/F at x as a compact linear group. It therefore leaves invariant some non-singular quadratic form. If any such invariant form is selected at the origin it may be transported to each point of M in a unique way because of the invariance. In this way we see that G/F has a Riemannian metric invariant under the action of G on G/F. We come now to the theorem about neighboring subgroups (Montgomery-Zippin [8]).

THEOREM. *Let G be a Lie group and F a compact subgroup of G.*
Then there exists an open set O in G, F ⊂ O, with the property that
if H is a compact subgroup of G and H ⊂ O, then there is a g in G
such that

$$g^{-1}Hg \subset F.$$

Moreover given any neighborhood W of e, O can be so chosen that
for every H ⊂ O the desired g can be selected in W.

Let $G/F = M$. The group G acts transitively on M and since
F is compact the space M as mentioned above may be given a
Riemannian metric which is invariant under the natural action
of the group G. If x is any point of M then for sufficiently small
neighborhoods U of x the topology of U is equivalent to the
topology given by the *geodesic* metric $d(y, z)$ where for y, z in U,
$d(y, z)$ is the length of the shortest geodesic joining y and z. In
the course of the proof of the theorem attention will be restricted
to neighborhoods of this kind and it will be assumed that they are
metrized by the geodesic metric as mentioned.

The symbol $S(x, r)$ denotes the closed set of points y whose
distance from x is at most r. The boundary of this sphere is the
set y of points whose distance to x is precisely r. A sphere is
called *convex* if each two points of the sphere are the end points
of a geodesic arc every inner point of which is in the interior of
the sphere. It should be remembered that if a geodesic arc is
in a convex sphere, then only the end points can be on the boundary
of the sphere.

The proof of the theorem will be based on a series of simple
lemmas all having the same hypothesis as the theorem; x is arbitrary but a definite point.

LEMMA 1. *If x is any point of M there exists a neighbourhood U*
of x such that if a, b, c are any three points of U which lie on a
geodesic in that order, then

$$d(b, x) < \max [d(a, x), \ d(c, x)].$$

Let r be a positive number such that every sphere of center
x and radius $r' \leqq r$ is convex. The existence of such a number r

has been proved by Whitehead [1]. Let U be the interior of $S(x, r)$, and let a, b, c be points of U which lie in that order on a geodesic arc abc in U.

Consider the sphere $S(x, r_1)$ where $r_1 = d(x, b)$. The boundary of this sphere meets the geodesic arc abc at b and this means that either a or c is outside $S(x, r_1)$ as otherwise $S(x, r_1)$ would not be convex. But if a point is outside $S(x, r_1)$ its distance from x is greater than r_1 and this completes the proof of the lemma.

LEMMA 2. *If x is any point of M there exists a neighborhood U of x such that if a, b, c, and u are four distinct points in U and a, b, and c are on a geodesic arc in U in that order then*

$$d(b, u) < \max\,[d(a, u),\ d(c, u)].$$

For the proof let $U = S(x, r/2)$, where r has the same meaning as in the preceding lemma. There is an element g in G such that

$$g(u) = x.$$

The points $g(a)$, $g(b)$, $g(c)$ are on a geodesic arc inside $S(x, r)$ because g is an isometry (since we introduced an invariant metric) and preserves distance and geodesic. Making use of Lemma 1 and again using the fact that g is an isometry completes the proof of the lemma.

LEMMA 3. *If x is a point of M there is a neighbourhood U of x such that if a, b, c are distinct points lying in that order on a geodesic arc in U and if d, u, f are distinct and lie in that order on a geodesic arc in U, then*

$$d(b, u) < \max\,[d(a, d),\ d(a, f),\ d(c, d),\ d(c, f)].$$

By Lemma 2, $d(b, u)$ is either less than $d(a, u)$ or less than $d(c, u)$, say the notation is such that

$$d(b, u) < d(a, u).$$

Using Lemma 2 again gives

$$d(a, u) < \max\,[d(a, d),\ d(a, f)]$$

and this completes the proof of the lemma.

LEMMA 4. *If x is any point in M, there exists an arbitrarily small neighborhood U of x such that if H is a compact subgroup of G and $H(x)$ is in U then for some b in U, $H(b) = b$, that is b is a stationary point of H.*

Let U denote the interior of some convex sphere about x which satisfies the conclusion of Lemma 3. It will be shown that this U also satisfies the present lemma. In order to carry out the proof assume that $H(x) \subset U \subset \overline{U}$.

Choose $q \in \overline{U}$ such that $H(q)$ is of minimum diameter among those orbits of H which lie entirely in $\overline{U}$. If $H(q) = q$ there is nothing more to prove so it may be assumed that $H(q) \neq q$.

Therefore $H(q)$ contains more than one point. Let b denote the midpoint of a geodesic arc in $\overline{U}$ whose endpoints are in $H(q)$. Since H is a group of isometries, $H(b)$ is entirely included in the set of such midpoints, and $H(b)$ is in U. It will be shown next that

$$\text{diam } H(b) < \text{diam } H(q)$$

which will be a contradiction to the choice of q, and this contradiction will prove that $H(q) = q$.

There is an h in H (since H is compact) such that

$$\text{diam } H(b) = d[b, \, h(b)].$$

The point b is the midpoint of a geodesic arc abc where a and c are in $H(q)$. The point $h(b)$ is the midpoint of the geodesic through $h(a), h(b), h(c)$. Denote these points by d, u, f and apply Lemma 3 to obtain

$$d[b, \, h(b)] < \max \{d[a, \, h(a)], \ d[a, \, h(c)], \ d[c, \, h(a)], \ d[c, \, h(c)]\}.$$

This shows that the diameter of $H(b)$ is smaller than that of $H(q)$ and this contradiction proves the lemma.

By using these lemmas the proof of the theorem will now be concluded. The points of M are images of G under the natural map

$$T : g \rightarrow gF.$$

Let $T(F) = x$ and let U be as in Lemma 4. Define

$$O = T^{-1}(U)$$

so that O is an open set in G which includes F. Let H be a compact subgroup of G, $H \subset O$. Then $T(H) = H(x) \subset U$ and by Lemma 4 some point of U is a stationary point of H. This means that there is a coset $gF \subset O$ such that

$$HgF = gF.$$

Hence

$$g^{-1}HgF = F$$
$$g^{-1}Hg \subset F$$

and the first statement of the theorem is proved.

The last statement is a consequence of the method of proof above.

Thus the set $T(W)$ is a neighborhood of x. If the neighborhood U of the last lemma is small then H will have a stationary point y near x and by choosing U sufficiently small it may be arranged that

$$y \in T(W).$$

Then there is a $g \in W$ such that $T(g) = y$. Hence $HgF = gF$, $g^{-1}Hg \subset F$ as before.

5.4. Cross-sections

It is not possible in general to find a closed set which forms a cross-section (that is a one-one image) of a *family* of orbits.

One sees this readily in the following simple example of cosets: let H be the subgroup of a circle group C consisting of the two elements of order two, and let H act on C by translation.

But under reasonably general circumstances there do exist *local* cross-sections. This was shown for one-parameter groups by Whitney [1] and is easily extended to abelian local n-parameter groups. The following more inclusive theorem is due to Gleason [2]. The theorem is proved here for *compact* Lie groups, but it can be proved for local Lie groups (therefore including the one-parameter groups).

THEOREM 1. *Let G be a compact Lie group which acts as a trans-*

formation group on a Hausdorff space M and suppose that a point $p \, \epsilon \, M$ *is in a neighborhood* X *in* M *with the property that if* $x \, \epsilon \, X$ *and* $g_1, \, g_2 \, \epsilon \, G$ *then* $g_1(x) = g_2(x)$ *only if* $g_1 = g_2$. *Then the orbits of* G *have a local cross-section at* p.

It will be proved that there is a compact set F including p such that for every x in some neighborhood of p there is a unique f in F and a unique r in G such that

$$x = r(f).$$

This is a little more than is required and shows that points on the orbits near p form a topological product of F and G.

Let H be a faithful representation (2.20) of G in the space of $n \times n$ matrices $Gl(n, R)$. If a and b are in G then $H(a)$, $H(b)$, $H(ab)$ are in $Gl(n, R)$ and

$$H(ab) = H(a)H(b).$$

The Lie group G possesses a left invariant measure.

A function $J(x)$ will be defined for $x \, \epsilon \, G(X)$ and with values in the space of $n \times n$ matrices M_n. On $G(p)$ let J be defined as follows

$$J(a(p)) = H(a) \quad \text{for } a \, \epsilon \, G$$

Now let J be extended to $G(X)$ in any continuous way, by Tietze's theorem (p. 80 of Hurewicz-Wallman).

Making use of these definitions

$$\int_G H(a^{-1})J(a(p))\,da = \int_G H(a^{-1})H(a)\,da$$
$$= \int_G I\,da, \quad (I = \text{unit matrix})$$
$$= AI$$

where A is a positive real number. Assume the measure in G normalized so that $A = 1$, and then with this convention

$$\int_G H(a^{-1})J(a(p))\,da = I.$$

Now define a continuous function $K(q)$ for q in $G(X)$ as follows:

$$K(q) = \int_G H(a^{-1})J(a(q))\,da.$$

By the remarks above

$$K(p) = I.$$

It will be convenient to prove for q in X, $r \in G$, that $K(rq) = H(r)K(q)$ and to do this proceed as follows (let $ar = b$, $a = br^{-1}$):

$$K(rq) = \int_G H(a^{-1})J(arq)da$$
$$= \int_G H(rb^{-1})J(bq)db$$
$$= H(r)\int_G H(b^{-1})J(bq)db$$
$$= H(r)K(q).$$

The group G acts not only on the space M but also on the space of $n \times n$ matrices through the representation H. It is known that local cross-sections exist for cosets of a subgroup in a Lie group.

There is a compact neighborhood N_1 of the identity matrix (in the set of $n \times n$ matrices) and a closed subset C_1 of N_1 such that if $m \in N_1$ there is exactly one point of C_1 of the form $H(r)m$, $r \in G$. It may be assumed that N_1 contains only non-singular matrices. Let $N = K^{-1}(N_1)$ and $C = K^{-1}(C_1)$. Then N is a compact neighborhood of p and C is a compact subset of N.

Suppose x is in N. Then $K(x)$ is in N_1 and there is exactly one point of C_1 of the form

$$H(r)K(x).$$

Since $K(x)$ is non-singular the element of C_1 can be so represented in only one manner. Therefore there is precisely one $r \in G$ such that

$$K(r(x)) = H(r)K(x)$$

is in C_1. Therefore there is exactly one $r \in G$ such that $r(x)$ is in C. This completes the proof of the theorem.

THEOREM 2. *Let R be a compact Lie subgroup of a locally compact group G. Then there is a compact local cross-section at e for the cosets of R.*

This is a corollary of the theorem above because the cosets can be regarded as orbits under R acting by translation in G.

THEOREM 3. *Let G be a compact Lie group which acts as a transformation group on a Hausdorff space M. Suppose $p \in M$ and for x in some neighborhood of p that G_x is conjugate to G_p. Then the orbits of G have a local cross-section at p.*

Since the interest is in local properties there is no loss of generality in assuming that for every x, G_x is conjugate to G_p. This assumption having been made let A be the set of points each of which is left stationary by all of G_p. Note that $p \epsilon A$ and that A touches every orbit.

Let N be the normalizer in G of the subgroup G_p. Then A is invariant under N. Since every point of A is left fixed by G_p, it is possible to define in a natural way the action of N/G_p as a transformation group on A. Thus let gG_p be any coset in N/G_p and define for $x \epsilon A$,

$$(gG_p)(x) = g(x).$$

This group N/G_p thus acts so that e is the only one of its elements which has a fixed point. By the theorem above the orbits of N/G_p have a local cross-section F at p. Since within A there is one and only one orbit of N/G_p in each orbit of G, the cross-section F is also a local cross-section for the orbits of G. This completes the proof.

REMARK. Let G be a compact Lie group acting on a locally compact space M. Let k be the highest dimension of any orbit of G in M and let O be the set of all points of M on k-dimensional orbits. Then O is an open set. If x and y are any two points of O then $\dim G_x = \dim G_y$. In O choose a point p where G_p has as few components as possible. By a theorem above (5.3) it follows for all x near p that G_x is conjugate to a subgroup of G_p. But since $\dim G_x = \dim G_p$ and since G_p and G_x have the same number of components it follows that: *For all x near p, G_x is conjugate to G_p. Therefore the orbits of G have a local cross-section at p.*

5.5. Periodic and pointwise periodic homeomorphisms

Among compact groups of transformations, finite groups and especially finite cyclic groups are a very important class. They are of great interest in themselves and give information about compact Lie groups of transformations since in a compact Lie group the periodic elements form an everywhere dense set. We first state without proofs a few of the interesting results about periodic transformations, and later use one of these results to prove a theorem about pointwise periodic homeomorphisms.

The following two theorems were first proved by Newman [1] and later extended and sharpened by P. A. Smith [2]. We formulate them for manifolds although Smith has shown them to be true for spaces which resemble manifolds in homology properties but may be more general. Recall that a manifold is a connected locally euclidean space.

THEOREM 1. *Let T be a periodic homeomorphism acting on a manifold M. If the set of fixed points of T contains an inner point then every point is fixed and T is the identity.*

For simplicity we formulate the following theorem in terms of a metric although it can be formulated in terms of coverings without a metric.

THEOREM 2. *Let M be a metric manifold. Then there exists a positive number r depending only on M such that if T (not the identity) is any periodic homeomorphism of M then some point has an orbit whose diameter is greater than r.*

The following theorem was proved by Smith [1] for euclidean spaces and more generally for spaces with homology properties similar to euclidean spaces.

THEOREM 3. *Let E be euclidean n-space and T a periodic homeomorphism of E of prime power period. Then T has a fixed point.*

Smith has shown that when T acts on M with period p and if M resembles a sphere in mod p homology then the set of fixed points F resembles mod p a sphere of some lower dimension. In the special case where $M = S^3$ then Smith has shown that F is either S^0 (that is a pair of points) S^1 or S^2. Many other interesting results have been given by Smith. Recently Floyd and Liao have extended our information about these questions.

Bing [1] has given an example of a period two homeomorphism T of S^3 onto itself where the set of fixed points F is homeomorphic to S^2 but where S^2 is not *tamely imbedded* (see 6.0). Hence T cannot be equivalent to an orthogonal transformation. His example can be modified to give a period two homeomorphism of S^3 where F is a simple closed curve which is not tamely imbedded.

A homeomorphism T of a space M onto itself is called *pointwise periodic* if for every x in M there is an integer k depending on x such that $T^k(x) = x$. If there exists a k such that $T^k(x) = x$ for all x then of course T is periodic. In general there are pointwise periodic homeomorphisms which are not periodic but the following shows (Montgomery [1]) that this cannot happen for well behaved spaces.

THEOREM. *Let M be a connected locally euclidean space and let T be a pointwise periodic homeomorphism of M onto itself. Then T is periodic.*

Let T^0 be the identity and for each x let $p(x)$ be the *period* of x, that is by definition $p(x)$ is the smallest positive integer k such that $T^k(x) = x$. The function $p(x)$ is lower semi-continuous.

Let K be the collection of points x such that on every neighborhood of x, $p(x)$ has an infinite least upper bound. Then K is a closed subset of M and is nowhere dense because $p(x)$ has finite values and as a semi-continuous function its points of continuity are everywhere dense. The proof of the theorem will depend on the following.

LEMMA. *On every connected open set R in $M - K$, the function $p(x)$ is bounded.*

Let C be the set of points of continuity of $p(x)$. As already remarked this set is everywhere dense in M. In some neighborhood of each point of C the function $p(x)$ is constant, and C is an open set.

Therefore there is a connected open set H in R such that $p(x)$ is constant and equal to an integer k everywhere in H. If H is not all of R let b be a point of R on the boundary of H. There is an open neighborhood U of b in R and an integer N such that for any point x in U, $p(x)$ is at most N. This is because $p(x)$ has a finite least upper bound at b.

Consider the homeomorphism $S = T^k$ and let

$$H_1 = \cup_{i=0}^{(kN)!} S^i[H \cup U].$$

This set is transformed onto itself homeomorphically by S, because any point y of $H \cup U$ has period at most N under T and therefore y has period at most kN under S.

The homeomorphism S, on H_1, is periodic of period at most $(kN)!$ and for x in the open subset H of the connected locally euclidean set H_1, it is true that $S(x) = x$. But by theorem 1 (Newman) it then follows that every point in H_1 is fixed under S. This means that if x is in H_1, then $T^k(x) = x$ so that every point of H_1 has period at most k under T.

This shows that if H is not all of R then H may be enlarged to a connected open set H_1 in R such that every point of H_1 has period at most k. If H_1 is not all of R then it may be enlarged by a similar process to a connected open set H_2. Since the union of open sets is open we may continue by transfinite induction if necessary and finally obtain a set $H_\alpha = R$. Then every point of H_α has period at most k. This completes the proof of the lemma.

The theorem will now be proved by the method of contradiction. Assume that $p(x)$ is unbounded on M. On the basis of this assumption the lemma above shows that $M - K$ is not connected and therefore that the closed set K is not vacuous.

Let $p(x/K)$ denote $p(x)$ restricted to K; $p(x/K)$ is semi-continuous on K. Let a be a point of K where $p(x/K)$ is continuous. There is an open set $H \subset M$ including a such that $p(x/K)$ is constant and equal to some integer k everywhere in $H \cap K$. Let R be any component of $M - K$ having points in H. There is by the lemma an integer N such that every point of R has period at most N. Let $S = T^k$ and define

$$W = \cup_{i=0}^{N!} S^i(R).$$

If any point of $\overline{R} \cap K \cap H$ is an inner point of $W \cup K$, $p(x)$ is bounded at that point. Let c be any point of $\overline{R} \cap K \cap H$. This point is in K and by definition of K and the remark just made, the point c cannot be an inner point of $W \cup K$. Let V be a connected open set which is in H and which contains c. The set $V - \overline{W}$ is not vacuous. Furthermore $V - \overline{W}$ is separated from W by K and each point of $K \cap H$ is fixed under S so that the transformation

$S*$ now to be defined is a homeomorphism of $V \cup W$ onto itself. The definition of $S*$ is as follows:

$$\text{for } x \in W, \; S*(x) = S(x)$$
$$\text{for } x \in V - W, \; S*(x) = x.$$

The set $V \cup W$ is a connected locally euclidean space on which $S*$ is a periodic homeomorphism whose fixed point set includes an interior point. Consequently $S*$ leaves fixed all of $V \cup W$.

Therefore every point of W and in particular every point of R has period at most k. Since R is an arbitrary component of $M - K$ with points in H, it follows that every point of $(M - K) \cap H$ has period at most k. This is also true for every point of $H \cap K$. Therefore at any point of $H \cap K$, $p(x)$ has a finite least upper bound. This contradicts the definition of points of K and proves that K is vacuous. This completes the proof of the theorem.

5.6. Simply connected homogeneous spaces

If G is a Lie group and H is a closed subgroup then a great deal is known about the homogeneous space G/H, especially when G is compact (the case G non-compact is apparently much more difficult). In this direction many contributions have been made by A. Borel, E. and H. Cartan, C. Chevalley, G. Hirsch, H. Hopf, G. D. Mostow, H. Samelson, H. C. Wang and others (see Samelson [2]). We do not attempt to discuss this interesting topic but we do include a theorem (Montgomery [12]) which has a rather simple geometric proof involving some facts about fibre mappings.

THEOREM. *If G is a connected Lie group which acts transitively on a compact simply connected manifold M, then G contains a compact subgroup which also acts transitively on M.*

The proof depends on the fact already mentioned (4.13) that if G is a connected Lie group then G is the direct product, as a space, of a maximal compact subgroup K and a euclidean space E. There always exist maximal compact subgroups and any two such are conjugate. Any given compact subgroup is in a maximal compact subgroup. In view of these facts let the group G of the theorem above be the direct product, as a space, of K and E

$$G = K \cdot E.$$

Since $M = G/G_x$ is simply connected, G_x must be connected. Since G_x is connected, G_x is the direct product of a compact subgroup L and a euclidean space F

$$G_x = L \cdot F.$$

It may be assumed that $L \subset K$ by the remarks above.

Let T_1 be the map from G/L to G/K under which each coset of L is carried to the coset of K which contains it.

$$T_1 : G/L \to G/K.$$

This is a fibre mapping in the sense that each point of G/K is in a neighborhood whose inverse is a product of a fibre and a local cross section. Each fibre is homeomorphic to K/L. But G/K is homeomorphic to E, a euclidean space of some dimension, and when a fibering has such a base space, it is a product (proved by Seifert in two dimensions and by Feldbau in general). Therefore G/L is homeomorphic to the topological product of K/L and E,

(1) $G/L = K/L \times E.$

Let T_2 be the map from G/L to $G/G_x = M$ under which each coset of L is carried to the coset of G_x in which it is contained.

$$T_2 : G/L \to G/G_x.$$

This is a fibre mapping and each of the fibres is homeomorphic to G_x/L, that is, to F, a euclidean space. But when all the fibres are homeomorphic to euclidean space it is known that a cross section A exists in the large. The set A is a closed set in G/L, touching each fibre of T_2 precisely once, and A is homeomorphic to the base space $G/G_x = M$.

There exists a mapping P which *retracts* G/L onto A;

$$P : G/L \to A,$$

that is, P maps all of G/L into A and each point in A is fixed under P. If a cycle in A bounds in G/L, it also bounds in A. The set A

is a manifold and hence contains a non-bounding cycle z, mod 2, such that

$$\dim z = \dim A.$$

Since z does not bound in A, the map P shows that z does not bound in G/L.

From (1) it follows that the homology properties of G/L are the same as those of K/L, and therefore $\dim M = \dim A = \dim z \leqq \dim K/L$.

Since L is a maximal compact group in G_x, we have

$$L = K \cap G_x.$$

Then $K(x)$ is homeomorphic to K/L

$$K(x) = K/L$$

so that

$$\dim M \geqq \dim K(x) = \dim K/L \geqq \dim M.$$

Since $K(x)$ is a manifold of the same dimension as M,

$$K(x) = M$$

and this completes the proof of the theorem.

CHAPTER VI

Compact Transformation Groups

6.0. Introduction

In the last sections of this chapter we shall report in some detail on compact connected transformation groups of three-space (and of lower-dimensional euclidean spaces). The results here are as favorable as one could wish — the action of such transformation groups on three-space being topologically indistinguishable from that of subgroups of the orthogonal group.

The situation is more complex for groups that are *not* connected, and for groups on higher dimensional spaces. Thus we have mentioned the example by Bing [1] of a "quasi-reflection" in three-space. In this involutory homeomorphism of S^3 the set of fixed points is a topological plane but the two complementary domains are not homeomorphic to three-cells (the plane is *wildly-imbedded*). Such a transformation cannot be linear in any coordinates. A simple modification of this example gives another example of a sense-preserving involution of three-space which has a wildly-imbedded topological line of fixed points; and using that modification the authors have constructed an example of a circle group acting in four-space E_4 [14] so the action cannot be differentiable in any differentiable structure of E_4. Yet even in this example the behavior on the whole shows many regularities and considerable resemblance to the case of a linear group of transformations. The extent to which this resemblance is valid in general is an interesting topic for study. There are indications that the possibility of irregular behavior of a compact transformation group is associated with the absence of orbits of high dimensions. It thus becomes a plausible program to discover what can be said about the action of a compact connected group on a manifold; for example, one may ask what the structure of the fixed point

[229]

set is like, what different types of orbits can occur, how the orbits are related to each other and to the fixed point set. Other questions worth attention are suggested in the preceding chapter as well as this one.

In the first sections of this Chapter we include some theorems of a general nature which are intended as the groundwork of such a research program supplementing the work of the preceding Chapter. A special case of considerable interest (when an $n-1$ dimensional orbit is present in n-space) is worked out in considerable detail in 6.5.

We shall confine ourselves to groups which are separable metric as well as locally compact (in this connection, compare 2.11 in Chapter II). This is to be understood in the sequel.

Some of what is proved is applicable also to locally compact groups but it is clear from an example below that the situations are more complicated if the condition of compactness is dropped.

Let E denote the real line, let C denote a circle group given to us as a transformation group of E. Then each orbit $C(x)$, $x \varepsilon E$, is a point, or a circle, and it cannot be the latter. Therefore the action of C is trivial: every point is a fixed point. Since the line has no compact connected subsets which are homogeneous, every compact connected group would give the same result. On the other hand, every open interval is homogeneous — and the group of similitudes Q (see Chapter II, 2.14, example 4) as well as the vector group V_1 can act on it leaving the endpoints of the interval fixed.

In the plane, and in higher dimensional spaces too, the group V_1 (for example) can operate effectively in one region of space and leave another region pointwise fixed. This makes it possible to have the group V_1 operate so that its action in one part of space is not influenced by its action in a different part. This also makes it possible to have a product-group operate so that each factor acts more or less independently of the others.

It is not known whether the same phenomenon can occur for the action of compact groups (it cannot occur for compact *Lie* groups); in particular it is not known whether the p-adic groups can operate effectively on euclidean spaces. *If not*, then it would follow that every locally compact group acting effectively on a manifold (locally-euclidean, connected) was necessarily a Lie group.

However, when a locally compact group is *transitive* on a manifold then its action is more amenable to study. This case is of especial importance in connection with the use of transformation groups in geometry. We conclude the book with a brief mention of only one aspect of this problem, having to do with the foundations of geometry of three-space along lines which Hilbert [2] developed for plane geometry.

6.1. Cross-sections and zero-dimensional groups

In this section we first prove a general theorem which pertains to finding a cross-section to an arc of sets, each set being compact and totally disconnected. The theorem, proved by Whyburn [1] and by the authors [2], is related to the subject of covering spaces and we make some application of it along those lines.

THEOREM 1. *Let M and N be compact metric spaces and let f be an open continuous map from M to N whose inverses are totally disconnected. Let $\alpha^*(t)$ be a map of the unit interval R_1, $0 \leq t \leq 1$, into N. Then if $f(a) = a^* = \alpha^*(0)$, there is a map $\alpha(t)$ of R_1 into M such that*

$$\alpha(0) = a, \quad f\alpha(t) = \alpha^*(t).$$

For each $n \, \varepsilon \, I$, let D_n be the set of dyadic rationals of the form:

$$i/2^n, \ 0 \leq i \leq 2^n$$

and let $D = \cup D_n$. Let $\beta_n(t)$ be a function defined on D_n to M such that $\beta_n(0) = a$ and if $t \, \epsilon \, D_n$, $f\beta_n(t) = \alpha^*(t)$ and subject to the further condition

$$d[\beta_n(i/2^n), \beta_n(i + 1/2^n)] = \text{dist} \ [\beta_n(i/2^n), f^{-1}\alpha^*(i + 1/2^n)].$$

From the sequence of functions $\beta_n(t)$ it is possible to select, by the diagonal process, a subsequence $\beta_{n_i}(t)$ which is convergent on all dyadic rationals D (of course for a fixed t in D there may be a finite number of n_i's for which $\beta_{n_i}(t)$ is undefined). Denote the limit on D by $\alpha(t)$.

The function $\alpha(t)$ will now be shown to be uniformly continuous on D. Assume that this is not true and that there is an $r > 0$ such that for every k there are dyadic rationals s_k and t_k satisfying

$$| s_k - t_k | < 1/k, \quad d[\alpha(s_k), \alpha(t_k)] > r.$$

There is no loss in assuming that s_k and t_k converge to a real number t.

Let $\epsilon > 0$ be given. When $n_i > K$, K some integer, it is true that the distance between two successive functional values of β_{n_i} is less than ϵ. Using this it follows that between s_k and t_k there is a set $x_1 = s_k, \ldots, x_{n'} = t_k$ of increasing dyadic rationals, satisfying

$$d[\beta_n(x_j), \beta_n(x_{j+1})] < \epsilon$$
$$d[\beta_n(x_1), \beta_n(x_{n'})] > r.$$

These functional values form an ϵ-chain of diameter greater than r. Hence $f^{-1}(t)$ must contain a connected set of diameter at least r but this is impossible. Hence $\alpha(t)$ is uniformly continuous.

Therefore $\alpha(t)$ may be extended so as to be defined and continuous on all of R_1 and this proves the theorem.

6.1.1.   Let G be a compact group which acts as a transformation group on a *normal* (1.24) Hausdorff space M.

The *orbits* of G denoted by $G(x)$, $x \in M$, are mutually disjoined compact sets which form a collection denoted by M^* and also by M/G. There is a map

$$T : M \to M^*$$

under which a point of M is mapped on the orbit containing it. The collection M^* is made into a space, also denoted by M^*, by defining a set W^* in M^* as *open* if $T^{-1}(W^*)$ is an open set in M. By its definition, T is a continuous map.

THEOREM 2.   *The space M^* defined above is a Hausdorff space and the map T is open. If M is metric so is M^*.*

Let Q be any open set in M. Then $G(Q)$ is a union of open sets and is also open. Next,

$$T(Q) = TG(Q) = Q^*.$$

However, Q^* is open because $T^{-1}(Q^*) = G(Q)$ is open; and therefore T is open.

Let a^* and b^* be two distinct points in M^* and let a and b be such that

$$a \in T^{-1}(a^*), \ b \in T^{-1}(b^*).$$

Let U and V be disjoined open sets including $G(a)$ and $G(b)$ respectively. Choose X and Y to be open sets such that

$$a \in X, \ G(X) \subset U$$
$$b \in Y, \ G(Y) \subset V,$$

as is possible by the compactness of G. Then

$$TG(X) \cap TG(Y) = \Phi$$

and thus a^* and b^* are in open disjoint sets proving that M^* is a Hausdorff space.

If M is metric, then M^* can be metrized by using the Hausdorff distance (1.10.5) for orbits in M.

THEOREM 3. *Let G be a locally compact group acting effectively on the line or the circle. Then G is a Lie group.*

A theorem almost as general as this was proved by Brouwer in [1] in 1912, by methods which used the explicit construction of the coordinate systems appropriate to all the possible Lie groups of transformations of the line or circle.

Let J denote an arc on the line or circle and let the five distinct points p, q, r, s, t lie on J in the indicated order. There is a neighborhood $U \subset G$ of the identity such that for all $h \in U$, $h(q)$ is an inner point of the arc pqr and $h(s)$ is an inner point of the arc rst.

There is an open subgroup G' of G and there is a compact invariant subgroup $H \subset U$ such that G'/H is a Lie group. It is not hard to see that H is the identity. For, since H is compact, $H(q)$ has first and last points on pqr and $H(s)$ is in rst. Then one sees first that no element of H can be sense-reversing.

Let q_1 and q_2 be the first and last points of $H(q)$ in an ordering on the arc. Now let h be any element of H. If $h(q_2) \neq q_2$ then $h(q_2)$ is to the left of q_2. There is an x such that $h(x) = q_2$ and x must be to the right of q_2 as otherwise sense would be reversed by h. But then $h^{-1}(q_2) = x$ and the orbit $H(q)$ contains a point to the right of q_2 which contradicts the choice of q_2. Hence for all $h \in H$

$$h(q_2) = q_2 = q.$$

Hence $H(q) = q$ and $H(s) = s$. It follows by the same considerations that $H(x) = x$ for every x of the line or circle. Since G is effective this means that H is e, as asserted. Therefore the open subgroup G' is a Lie group and consequently G is a Lie group.

6.1.2. A pair (X, f) consisting of an arcwise connected space X and a map f of X is called a *covering space* of a space Y (compare 1.27.2) provided that (f is continuous)

1) $f(X) = Y$,

2) every point y in Y is in an open set V such that $f^{-1}(V)$ is the union of disjoined open sets each of which is mapped homeomorphically onto V by f.

A *path* $A(t)$ in a space is the continuous image of the closed unit interval $[0, 1]$. A path is called *closed* if the *ends* $A(0)$ and $A(1)$ are equal. Two paths $A(t)$ and $B(t)$ with the same ends are *homotopic* if there is a map $Q(s, t)$ of the rectangle $0 \leq s \leq 1$, $0 \leq t \leq 1$ such that

$$Q(0, t) = A(t), \quad Q(1, t) = B(t),$$
$$Q(s, 0) = A(0) = B(0), \quad Q(s, 1) = A(1) = B(1).$$

An arcwise connected space is called *simply-connected* if every closed path in the space is homotopic to a point. In such a space any two paths with the same ends are homotopic.

These remarks can all be applied to topological groups. A pair (G, f) consisting of an arcwise connected group and a continuous homomorphism f of G is called a *covering group* of a topological group H provided the conditions 1) and 2) above are satisfied for $G = X$ and $H = Y$.

The following theorem expresses one of the useful properties of covering spaces.

THEOREM 4. *Let (X, f) be a covering space of Y and let M be a simply connected space, containing a point m_0. If D^* is a map of M into Y taking m_0 into y_0 and if x_0 satisfies $f(x_0) = y_0$, then there is one and only one map D of M into X satisfying*

$$D(m_0) = x_0, \quad fD(m) = D^*(m), \quad m \in M.$$

The proof may be outlined as follows: It is first proved that the result is true when M is the closed unit interval or when it is a rectangle and m_0 is an end point (or a corner point). Next if m is any point in M let $A_1(t)$ be a path joining m_0 and m, and let $B_1(t)$ be the unique covering path of $D^*A_1(t)$ beginning at x_0. If $A_2(t)$ is a second path joining m_0 and m, then $B_1(1) = B_2(1)$. Hence it is permissible to define $D(m)$ to be $B_1(1)$. This definition gives a continuous map with the desired properties. The fact that it is unique follows from the same property for intervals.

As a Corollary to the foregoing one obtains:

Let (X, f) cover Y and let $A(t)$ be a path in X. If $fA(t)$ is a closed path in Y and can be shrunk to a point then $A(t)$ is a closed path in X and can be shrunk to a point.

6.1.3.   A transformation group is said to operate *without fixed points* if no element except the identity leaves a point fixed. This is the case, for example if a subgroup of a group operates on the group by translation (1.26.2).

THEOREM 5. *Let G be a compact zero-dimensional group acting without fixed points on a normal arcwise connected Hausdorff space X. If H is a compact open invariant subgroup of G then X/H and the natural map*

$$T : H(x) \to G(x)$$

is a covering space of X/G.

The proof is left to the reader.

By means of this theorem and the fact that compact zero-dimensional groups have arbitrarily small open and compact invariant subgroups it is possible to extend some of the facts about coverings mentioned above.

THEOREM 6. *Let G be a compact zero-dimensional group acting without fixed points on a normal arcwise connected Hausdorff space X and let f be the map taking X to the decomposition space X/G. Let M be a simply connected space containing a point m_0. If D^* is a map of M into X/G taking m_0 to y_0 and if $f(x_0) = y_0$ then there is one and only one map D of M into X satisfying*

$$D(m_0) = x_0, \quad fD(m) = D^*(m), \quad m \in M.$$

Let $H_1 \supset H_2 \supset \ldots$ be a sequence of compact invariant subgroups of G, intersecting in e, and such that each G/H_n is finite. Each space X/H_n is a finite covering X/G, and is also a finite covering of X/H_{n-1}. The space X is the inverse-limit of

$$X/G \leftarrow X/H_1 \leftarrow X/H_2 \leftarrow \ldots$$

and the theorem can be proved by successive "lifting" of the map D^* through the finite coverings.

In case M is metric a more direct proof can be made using Theorem 1. The details are left to the reader.

COROLLARY 1. *Let U be a neighborhood of x_0 in X. Then there is a neighborhood V of $f(x_0)$ in X/G such that if $D^*(M) \subset V$ then $D(M) \subset U$.*

A suitable neighborhood V can be obtained by the choice: $V = f(U)$. Verification of this, and of the succeeding corollaries is left to the reader.

COROLLARY 2. *An arc in X/G can be "raised" to an arc in X.*

COROLLARY 3. *If $A(t)$ is a path in X and if $fA(t)$ is a closed path in X/G which can be shrunk to a point then $A(t)$ is a closed path in X which can be shrunk to a point.*

6.2. Finite-dimensional groups and orbits

In this section we prove a few general Theorems about locally compact groups (*considered to be separable metric for simplicity*) acting transitively on a finite-dimensional set, or (more generally), acting on a space so that all orbits are of the same finite dimension. We begin by recalling the theorem which was proved in **4.9**.

THEOREM. *Every n-dimensional locally compact group G in some neighborhood U of the identity is the direct product of a compact zero-dimensional subgroup Z and a local n-dimensional Lie group R; thus*

$$U = Z \times R.$$

We may suppose that R is ruled by one-parameter local subgroups. If K is a compact invariant subgroup of G such that G/K is a Lie group, then dim $G/K \leqq n$.

The last remark in the theorem follows from the way in which the theorem was proved (namely the "lifting of n-cells").

The group G has open subgroups each of which is a projective limit of a sequence of Lie groups; and in fact the subgroup generated by R and Z is a typical group of this kind. For most of our purposes there would be no loss in supposing that G itself is generated by R and Z. In this case Z is a central subgroup of G, and every subgroup of Z is invariant.

Now let $Z_1 \supset Z_2 \supset \ldots$ be a sequence of compact invariant groups such that lim $Z_i = e$ and for each i, G/Z_i is Lie; let G_i denote G/Z_i. Then G_i is a Lie group and $G_1 \leftarrow G_2 \leftarrow \ldots$ is an inverse sequence whose projective limit is G. Any cell in G_i can be raised to a cell in G so dim $G_i \leqq$ dim $G = n$. Since the natural map of G onto G_i has inverse sets which are uniformly small (over any compact subset of G), it follows that dim $G_i \geqq n$, at least for large i (Alexandroff Approximation Theorem, see Hurewicz-Wallman). However, independently of this (from the way in which the previous theorem was proved) dim $G_i \leqq n$ for all i. There would be thus no loss of generality in assuming dim $G_i =$ dim G for all i, and that Z_{i-1}/Z_i is finite.

We turn now to the structure of coset spaces of a group G. Throughout this discussion *the group G is to be separable metric and locally compact.* Further we assume that G has arbitrarily small compact invariant subgroups giving Lie factor groups.

This means we assume the following:

A)   *G is the limit of a countable inverse sequence (G_i, f_i):*

$$G_1 \leftarrow G_2 \leftarrow \ldots$$

of Lie groups with f_i being a continuous homomorphism taking G_{i+1} into G_i; f_i has a compact kernel.

If G is compact or locally compact and G/G_0 compact it automatically satisfies A); in general G might not satisfy A) although it would always have an open subgroup satisfying A). If G is finite-dimensional it may be assumed that dim G = dim G_i for all i.

If H is a closed subgroup of G then associated with H, there is in each G_i a closed subgroup H_i, lim $H_i = H$. There is then a collection of coset spaces,

$$G_i/H_i$$

and the maps f_i induce maps of these coset spaces

$$f_i : g_{i+1}H_{i+1} \rightarrow f_i(g_{i+1}H_{i+1})$$
$$= f_i(g_{i+1})H_i.$$

Hence the coset spaces G_i/H_i and the induced maps form an inverse sequence. It will be left to the reader to prove that

$$G/H = \lim (G_i/H_i).$$

If dim G/H is finite it does not follow that dim G is finite. However it will be shown soon that if dim G/H is finite then some finite-dimensional group is transitive on G/H so that the structure of a finite-dimensional G/H will then be known to be the same as if G were finite-dimensional. We next consider the structure of G/H for a finite-dimensional G.

When G is finite dimensional we may assume

$$\text{dim } G_i = \text{dim } G$$
$$\text{dim } H_i = \text{dim } H$$

Each space G_{i+1}/H_{i+1} is a finite covering of the preceding and by the use of the theorems of 6.1. we can lift a cell from any one of these spaces up to the limit space G/H. We can also do this directly by the use of the first theorem of 6.1 because G/H is a "covering" of each G_i/H_i whose covering sets are compact and totally disconnected.

Now G_i is a Lie group and each G_i/H_i has the same dimension, say k,

$$\text{dim } G_i/H_i = \text{dim } G_i - \text{dim } H_i = \text{dim } G - \text{dim } H = k.$$

By the theorem of Alexandroff already mentioned it follows that $\dim G/H \leq k$. Hence $\dim G/H = k$.

6.2.1. The preceding discussion has been preliminary to the following important theorem on the local structure of n-dimensional orbits.

THEOREM. *Let G be an n-dimensional locally compact group which satisfies A) and let H be a closed m-dimensional subgroup. Then $\dim G/H = k = n - m$. The space of G/H is locally the topological product of a k-cell and a totally disconnected set D.*

It is in accord with the definition that a single point shall be called a totally disconnected set. However, if D is a single point (or a finite set) then G/H is locally euclidean; otherwise D is actually a perfect set.

The proof follows readily from the inverse limit sequence which was set up in the preceding section. Each G_i/H_i is k-dimensional, and carries a k-cell. This can be lifted to a k-cell in G/H. The desired k-cell can also be found by use of the local subgroup R. For H intersects R in a closed set, and there is a local Lie group R_1 which coincides with this intersection in a neighborhood of e. There is a k-cell $S_1 \subset R$ which is complementary to R_1 in the vector addition (3.9) and S_1 determines a k-cell of cosets of H. Let $Z_1 = H \cap Z$. The set D is determined by Z/Z_1.

6.2.2. The next theorem shows that a finite-dimensional G/H has the same structure that it would have for $\dim G$ finite.

If G is a locally compact group and H a closed subgroup then G acts transitively on the space $G/H = M$. If G contains a normal subgroup K which leaves all of M fixed then G/K can be defined in a natural way as a transformation group of M.

THEOREM. *Let G satisfy A). If H is a closed subgroup of G such that $\dim G/H = n$, then H contains a closed connected subgroup K such that a) K is invariant in G, b) G/K is finite dimensional. Thus if G acts on a finite dimensional coset space, then there is a finite dimensional group G/K which has the same coset space.*

If $\dim G/H = 0$, then K may be taken to be the identity-component $C_e(G)$. Hence it will be assumed that dim $G/H > 0$.

Let F_a be compact invariant subgroups of G such that $G/F_a = G_a$ is a Lie group, and let $K_a = C_e F_a$. The group G/K_a is mapped onto G/F_a with a compact zero-dimensional kernel. Since G/F_a is a Lie group it follows that dim $G/K_a = $ dim G/F_a which of course is finite. We suppose $F_1 \supset F_2 \supset \dots$, and that $\cap F_i = e$.

Each K_a acts on $G/H = M$ and since K_a may be selected in any preassigned neighborhood of e it follows that if $x \in M$ there is a value of a, say $a = N$ such that

$$K_N(x) \text{ is a proper subset of } C_x M.$$

Since K_N is invariant in G, the group G operates on the space, called $D(M, K_N)$ made up of the orbits $K_N(y)$, $y \in M$. In this operation K_N leaves every point of $D(M, K_N)$ fixed so in consequence there is a natural way in which

$$G/K_N \text{ operates on } D(M, K_N).$$

The space $D(M, K_N)$ has positive dimension because $K_N(x)$ is a proper subset of $C_x M$.

Let $x \in M$ map into x^* in $D(M, K_N)$.

Since G/K_N is finite dimensional it contains a local Lie group R such that dim G/K_N is dim R. Let G_{x^*} denote the subgroup of G/K_N which leaves x^* fixed. Then

$$R \cap G_{x^*}$$

is a local Lie subgroup of R. There is in R and therefore in G/K_N a local one-parameter group $h^*(t)$ such that for $|t| <$ some positive α, $h^*(t)$ meets G_{x^*} only in e. There is in G a one-parameter group $h(t)$ such that

$$h(t) \rightarrow h^*(t)$$

under the map

$$G \rightarrow G/K_N.$$

Hence $h(t)$ operates on x^* and if $t \neq t'$, $|t| + |t'| < \alpha$, then $h(t)(x^*) \neq h(t')(x^*)$. Now $h(t)$ operates on M.

It follows that

$$h(t)K_N(x) \text{ does not touch } h(t')K_N(x),$$

and for $|t| \leq \alpha$ that

$$h(t)K_N(x)$$

is homeomorphic to the product of $K_N(x)$ and an interval. But such a product has larger dimension than $K_N(x)$.

Therefore it has now been shown that

$$\dim K_N(x) < \dim M.$$

If $\dim K_N(x) > 0$ the same process may be applied to K_N and the orbit $K_N(x)$ gives an $n > N$ such that

$$\dim K_n(x) < \dim K_N(x).$$

Continuing a finite number of steps gives an index $i > N$ such that $\dim K_i(x) = 0$ and since K_i is connected,

$$K_i(x) = x.$$

But K_i is an invariant subgroup of G so that for all y in M, $K_i(y) = y$. It has already been remarked that $\dim G/K_i$ is finite, and the proof of the Theorem is completed.

6.2.3.   THEOREM. *Let G be a compact group which acts on a locally compact space M. Then the points of M on orbits of dimension $\geq k$ (k is any non-negative integer) form an open set.*

If the theorem is false, then M (using sequences for simplicity) contains x and a sequence $y_1, y_2, \ldots$, converging to x, such that for some pair of integers h and k, with $h < k$, and for all n,

$$\dim G(x) \geq k$$
$$\dim G(y_n) = h.$$

Set $F_n = G_{y_n}$ and set $H = G_x$. Every open subset of G containing H must contain all but a finite number of the sets F_n.

Let K denote any invariant subgroup of G such that G/K is a Lie group. Set $G' = G/K$ and let T denote the natural map of G onto G'. Set $F'_n = T(F_n)$ and set $H' = T(H)$. Every neighborhood of H' in G' contains all but a finite number of the groups F'_n and therefore, for sufficiently large n, F'_n is conjugate to a subgroup of H' (p. 216). For all such n there exists a map of G'/F'_n upon G'/H', and it follows that dim $G'/H' \leqq$ dim G'/F'_n.

For all n, there is natural map of G/F_n upon G'/F'_n, and it follows from dim $G/F_n = h$ that dim $G'/F'_n \leqq h$. Therefore, finally, dim $G'/H' \leqq h$.

It follows that G/H can be approximated by coset-spaces G'/H' of dimension not greater than h, and we get the contradiction

$$\dim G(x) = \dim G/H \leqq h < k.$$

6.2.4. Let G act transitively on a space M, and let K_M be the identity component of the subgroup of G which leaves all of M fixed. We showed that if M is finite dimensional there is a connected subgroup K leaving all of M fixed such that G/K is finite dimensional. Let us see that dim G/K_M is finite.

Now $K \subset K_M$ and $G/K_M = (G/K)/(K_M/K)$ and therefore dim $G/K_M \leqq$ dim G/K by 6.2.2. Many of the theorems of this section were proved by the authors in [6].

THEOREM. *Let G be a compact group acting on a locally compact space X. For all orbits N sufficiently near to an orbit M, K_N lies in a preassigned neighborhood V of K_M. If M is finite-dimensional then for N sufficiently near to M, $K_N \subset K_M$.*

The fact that K_N draws into a neighborhood V of K_M as N approaches M follows because otherwise there would be a connected set not in V and leaving all of M fixed.

Consider next the case where M is finite-dimensional and assume dim $M > 0$; for dim $M = 0$ the proof is immediate. Then G/K_M is also finite-dimensional. From the local structure of a finite-dimensional group there is an open set $U(e)$ in G/K_M which contains no non-trivial connected subgroup. Let T be the map

$$G \to G/K_M.$$

Then $T^{-1}(U) = V$ is an open set including K_M in G. Any connected group F in V must be in K_M for otherwise $T(F)$ would be a non-trivial connected group in U. This completes the proof.

6.2.5. THEOREM. *Let G be any effective compact Lie group of iso-metries of a connected analytic Riemannian manifold of dimension k. Then $\dim G \leq (1/2)k(k+1)$.*

The proof can be made by induction since G_x leaves invariant a small sphere S around x and must be effective on the manifold S; $\dim S = k - 1$. The details are left to the reader.

COROLLARY 1. *Let G be a compact connected Lie group effective on an orbit M whose dimension is k. Then $\dim G \leq (1/2)k(k+1)$.*

The Corollary follows because it is possible to introduce into M an invariant Riemannian metric.

COROLLARY 2. *The result of Corollary 1 holds for any compact connected G.*

This follows from Corollary 1 by a limit argument.

THEOREM. *Let G be a compact connected group acting effectively on a connected space X. If all orbits in X are of the same finite-dimension k then G is finite-dimensional.*

Let M be any orbit in X. Then G/K_M is finite-dimensional. For N near to M, $K_N \subset K_M$. Choose M so that G/K_M is of the highest possible dimension, which must be $\leq k(k+1)/2$. Now for N near to M, $K_N = K_M$. Then the set left fixed by K_M is open. But it is also closed and this proves the Theorem.

6.3. Orbits which are manifolds

LEMMA. *Let G be locally compact satisfying* A). *Let N_i, $N_i \supset N_{i+1}$, $\cap N_i = e$, be a countable collection of compact invariant subgroups such that $G/N_i = G_i$ (Lie group). If H is a closed subgroup such that G/H is finite-dimensional and locally connected then for some i, $N_i \subset H$.*

COROLLARY. *If a locally compact group satisfying* A) *is effective and transitive on a locally compact locally connected finite-dimensional space then G is a Lie group.*

Let $f_i : G/N_{i+1} \to G/N_i$ and for the proof recall that G/H is the limit of the system of spaces G_i/H_i and maps f_i (more exactly the maps induced by f_i which we shall also call f_i). Under the present hypothesis for large i, f_i must be a homeomorphism of G_{i+1}/Π_{i+1} onto G_i/Π_i, that is we have

  1) $f_i : G_{i+1} \twoheadrightarrow G_i$
  2) $f_i : G_{i+1}/H_{i+1} \twoheadrightarrow G_i/H_i$

and 2) is a homeomorphism. However 2) is a homeomorphism only if $N_{i+1} \subset H_{i+1}$. This completes the proof of the Lemma and Corollary.

6.3.1.   Suppose now that a compact metric group G acts on a manifold. Then as has been seen

$$G = \lim G_i$$

is the inverse limit of a countable sequence of Lie groups $G_i = G/N_i$ with maps $f_i : G_{i+1} \twoheadrightarrow G_i$.

If, further, G has a finite-dimensional locally connected orbit $G(x) = G/G_x$ then for some i, $N_i \subset G_x$.

If a compact metric group G as above acts on a space E in such a way that all orbits are locally connected and finite-dimensional, then there is defined on E a function $p(x)$ where $p(x)$ is the smallest integer p such that $N_p \subset G_x$. For the next Theorem, compare 5.5.

THEOREM 1. *If a compact group G acts effectively on a connected locally euclidean space E and if all the orbits in E are locally connected then G is a Lie group.*

As mentioned just above there is defined a function $p(x)$ everywhere in E. Let L be the set of points in E at which the least upper bound of $p(x)$ is infinite. The function is lower semi-continuous and is integral valued so that L is a closed nowhere dense set. The proof of the Theorem will make use of the following result for which we use a theorem of Newman [1].

LEMMA. *Let R be a component of $E - L$. Then $p(x)$ is bounded on R.*

It is clear that $p(x)$ is bounded in the vicinity of each point of R but the Lemma asserts further that there is a uniform bound for all of R.

Let O be a maximal open connected subset of R on which $p(x)$

is at most equal to k, when k is so chosen that O is not vacuous. Assume that O does not coincide with R, and let b be a point of R which is on the boundary of O and thus is not in O. Let U be a connected open neighborhood of b, $U \subset R$ such that $p(x)$ is bounded on U, say for $x \in U$, $p(x) \leqq r$, $k < r$.

The group N_k leaves each point of O stationary and this implies that

$$X = N_k(O \cup U)$$

is a connected locally euclidean space. The group N_r leaves every point of X stationary so that N_k/N_r may be defined as a transformation group of X in such a way that the orbits of N_k/N_r are the same as orbits of N_k.

Thus N_k/N_r is a compact Lie group acting on X and leaving O, an open set, stationary. This shows by a theorem of Newman that N_k/N_r leaves all of X stationary. The same is true for N_k. This contradicts the assumption that O was maximal, and this contradiction proves the Lemma.

The proof of the Theorem will now be made by using the Lemma.

Assume that $p(x)$ is not bounded on E. By the Lemma, L is not vacuous. Let $p(x \mid L)$ be the function $p(x)$ as restricted to L and let b be a point of L at which $p(x \mid L)$ is continuous. There is an open set O including b such that $p(x \mid L)$ is constant and equal to an integer k everywhere in $O \cap L$. Let R be any component of $E - L$ which has a point in O. There must be an integer r, $r \geqq k$, which bounds $p(x)$ in R. Let

$$X = N_k(R).$$

The function $p(x)$ is bounded on X and consequently no point of L can be an inner point of $\overline{X}$. Let c be a point of $\overline{X} \cap L \cap O$. This point cannot be an inner point of $\overline{X}$. Let $V \subset O$ be a connected open set which contains c. The set $V - \overline{X}$ is not vacuous and it is separated from X by L. For this reason it is possible to define N_k as a new transformation group of $X \cup V$ in a manner about to be described.

If x is in X and g is in N_k let

$$g\{x\} = g(x).$$

If x is in $V - X$, let

$$g\{x\} = x.$$

With the new action of N_k thus defined the group N_r leaves all of $X \cup V$ fixed so that the group N_k/N_r which is a Lie group acts in a natural way with the same orbits as N_k. But N_k/N_r leaves all of $V - X$ fixed and this includes an interior point. Hence N_k leaves all of $V \cup X$ fixed. Hence for $x \in V \cup X$, $p(x) \leqq k$.

The component R was an arbitrary one touching O and therefore for every point x of $O - L$, $p(x) \leqq k$. This shows that $p(x)$ is bounded at points of L in O and this contradicts the definition of L. This shows that L is vacuous and thereby completes the proof of the theorem.

THEOREM 2. *Let G be a compact connected Lie group which acts effectively on a connected locally euclidean space E and let k be the highest dimension of any orbit. Then $\dim G \leqq k(k + 1)/2$.*

This is known for G acting on an E consisting of a single orbit and this will be assumed. Consider now the more general case where G is not transitive on E. It will be shown that there is an x in E such that for $M = G(x)$ the group G_M is finite, where G_M is the group leaving all of M fixed.

Choose x so that G/G_M has as high dimension as possible. For orbits N near to M, $K_N \subset K_M$. The groups K_N and K_M are connected Lie groups; if $\dim K_N < \dim K_M$ then $\dim G/G_N > \dim G/G_M$ which is against the choice of M. Hence for N near M, $K_N = K_M$. Thus K_M leaves an open set fixed and therefore leaves all of E fixed. But G is effective so $K_M = e$.

Hence G_M is finite and $\dim G/G_M = \dim G$. The group G/G_M is effective on M, and by hypothesis $\dim M \leqq k$. The general case now follows from the special case of a single orbit and this completes the proof of the Theorem.

6.4. The orbit of an (n—1)-cell and application

THEOREM. *Let G be any compact group acting on a locally euclidean n-space E. Let K be a closed $(n - 1)$-cell in E and let C be*

the component of $G(K)$ which contains K. Then the set H of elements
h such that $h(K) \subset C$ is an open and closed subgroup of G. Also,
$G(K)$ has a finite number of components.

It will first be observed that H is a group. This is true because
C is invariant under H and in fact H could equally well be defined
as the set of all elements which leave C invariant. Furthermore
since C is closed, the subgroup H is closed.

It will be proved that H is open. Because H is a group it will
be sufficient to prove that the identity of G is an inner point of H.

6.4.1. The proof of this Theorem (Montgomery-Zippin [2]) rests
on the following topological fact which is a direct consequence of
fundamental separation properties of n-space (compare in Borsuk
[1], the notion of *free* subset; also in Wilder [1]). Let E be a locally-
euclidean n-space and let $S \subset E$ be a set which is the union of an
$(n - 1)$-cell and an arc one of whose endpoints is an inner point
of the cell, and the arc and cell have no other point in common.
Then there exists an $\epsilon > 0$ such that if f is a continuous map of
S into E and dist $(x, f(x)) < \epsilon$, then S and $f(S)$ have a point in
common. This may be expressed by saying that the set S is not a
free subset of E, or that S is not *free* in E.

Let O be an open n-cell, $\bar{O}$ compact, where O includes a point k on
the relative interior of K. Let K_1 be a closed $(n-1)$-cell interior to
K and in O and with k in the interior of K_1. Let B_1 be the boundary
of K_1. Let U be a neighborhood of e in G so small that $U(B_1)$ does
not separate between k and the boundary of O. Let $N \subset U$ be a
compact normal ($=$ invariant) subgroup of G so G/N is a Lie group.
Let $a_1 b_1$ be an arc drawn from $O - N(K_1)$ to $N(K_1) - N(B_1)$ so
that the only point of the arc in $N(K_1)$ is b_1, and so $N(a_1 b_1) \subset O$.

There is an element f in N such that $f(b_1)$ is in $K_1 - B_1$. The
arc $ab = f(a_1 b_1)$ is entirely in $O - N(K_1)$ except for the endpoint b
which is in K_1. For g in N and g sufficiently small it must be true
that $g(b)$ belongs to K_1 or that

$$(K_1 \cup ab) \cap g(K_1) \neq \Phi$$

or

$$K_1 \cap g(K_1 \cup ab) \neq \Phi.$$

This is impossible unless K and $g(K)$ intersect. Hence N contains a compact subgroup say N_1 such that G/N_1 is a Lie group and such that if $g \in N_1$ then $K \cap g(K) \neq \Phi$. Hence $N_1(K) \subset C$, and $N_1 \subset H$. It is also true that

$$G^* = C_e(G) \subset H$$

and therefore if F is the group generated by N_1 and G^*,

$$F \subset H.$$

However F must include e as an inner point in G. This completes the proof of the theorem.

6.4.2. THEOREM. *Let G be a compact group acting on a locally euclidean n-space E, and let K be an $(n-1)$-cell in E. Then $G(K)$ is locally connected.*

In order to prove that $G(K)$ is locally connected it is sufficient to prove that it is the union, for any ϵ, of a finite number of connected sets of diameter at most ϵ. The cell K may be expressed as the union of a finite number of cells K_i where each K_i has diameter less than $\epsilon/8$.

Let N be a compact normal subgroup of G such that G/N is a Lie group, and so that diam $N(K_i) < \epsilon/2$. The set $N(K_i)$ has a finite number of components by 6.4. Hence

$$N(K) = \cup N(K_i)$$

is the union of a finite number of connected sets of diameter $< \epsilon/2$.

The group G/N may be represented as the finite union of connected sets A_i^* each of diameter less than any desired positive number. Let A_i be the sets in G corresponding to the sets A_i^* in G/N. Then

$$A_i N(K_j)$$

is connected and $G(K)$ is represented as a finite union of these connected sets which can be so constructed as to have diameter at most ϵ. This completes the proof.

COROLLARY. *If a compact group G acts on a locally euclidean* *n-dimensional space E then any (n—1)-dimensional orbit is locally connected.*

This follows from 6.2.2 and 6.2.1.

THEOREM. *Let G be a compact zero-dimensional group which acts effectively on a two-dimensional connected locally euclidean space E. Then G is finite.*

The same fact for one-dimensional connected locally euclidean spaces has already been proved and will be used. Also $N(C')$, see below, is locally connected by the above.

Let D be a domain in E whose closure $\overline{D}$ is a two-cell and let D' be the boundary of D. Let C' be a simple closed curve in D and let C be the interior (in D) of C'. Let x be any point in C. There is a normal compact open subgroup N in G such that $N(x) \subset C$. Let K be a closed 2-cell in C such that

$$N(x) \subset K.$$

The group N may be chosen so that

$$N(C') \subset D, \ N(C') \cap K = \Phi.$$

The set $N(K)$ is a continuum such that

$$N(K) \cap N(C') = \Phi$$
$$N(K) \subset C.$$

The component of $E - N(C')$ which contains $N(K)$ is invariant under N. The boundary B' of this component is also invariant under N and separates x from points of D'. The component of $E - B'$ which contains D' is invariant under N. The boundary B'' of this component is invariant under N and separates x from D'.

It can be seen that B'', a subset of $N(C')$ is the common boundary of at least two domains, and must therefore be a simple closed curve (Whyburn [1], p. 106). By the theorem for one-dimensional manifolds there is a compact open subgroup N'' of N which leaves every point of B'' stationary. It will now be shown that N'' leaves every point z of E stationary.

There exists one arc azb such that a and b are on B'' and the arc azb separates the domain of $E - B''$ in which it lies. In the closure of this domain it separates a pair of points c and d in B''.

By a similar argument to the above $N''(azb)$ contains an arc which is invariant under N'' and which joins a and b. The points a and b are fixed under N'' and it follows that every point of the invariant arc is fixed under N''. It now follows that the invariant arc coincides with azb and hence that z is a fixed point. This completes the proof.

6.5. $(n-1)$-dimensional orbits in E_n

We now consider a compact group acting on E_n with at least one $(n-1)$-dimensional orbit. In this case the action of G can be rather completely analyzed as we shall indicate (see Montgomery-Zippin [4]). In all of Chapter 6, G is a separable metric.

THEOREM. *Let G be a compact connected group which acts on euclidean n-space E with at least one $(n-1)$-dimensional orbit. Then all orbits except one are $(n-1)$-dimensional.*

We know that $G(x)$ is a manifold and separates E into precisely two domains. If E^* is the decomposition space of E in which each orbit of E becomes a point, then the point x^* corresponding to x cuts E^* into precisely two connected sets. Thus x^* is a cut point and a cut point of order 2. The space E^* is such that the cut points form an open set and each cut point is of order 2. By cyclic element theory (see Whyburn [1]) E^* is an arc, a ray, or a line. It follows that E^* is a ray because a cut point of E^* divides E into two domains one of which has a compact closure. This completes the proof of the Theorem. If G is effective it is a Lie group.

6.5.1. THEOREM. *Let G be a compact connected group which acts on euclidean n-space E with at least one $(n-1)$-dimensional orbit. If x and y are not in the exceptional orbit of lower dimension then G_x is conjugate to G_y.*

Denote the identity components of G_x and G_y by G_x^* and G_y^*. It will be shown first that G_x^* and G_y^* are conjugate.

If z is on the exceptional orbit then for all x and y in $E - G(z)$ it is true that $\dim G_x^* = \dim G_y^*$. It is known that if y approaches x then G_y and G_y^* draw into any given neighborhood of G_x and G_x^*. Hence for y near x, G_y^* is conjugate to G_x^*.

Now let x be a definite point in $E - G(z)$. By the remarks above points y such that G_y^* is conjugate to G_x^* form an open set. However it is also true that this set is closed in $E - G(z)$ so that it must be equal $E - G(z)$. Therefore for any x and y in $E - G(z)$, G_x^* is conjugate to G_y^*.

Next consider the integral valued function $\alpha(x)$ defined for x in $E - G(z)$ and equal to the number of components in G_x. This function is constant on each orbit so that there is an associated function $\alpha^*(z^*)$ defined in $E^* - z^*$. Both of these functions are upper semi-continuous and this, together with the fact that they are integral valued implies that they are constant on certain open sets. If $\alpha(x)$ is constant everywhere in $E - G(z)$ then G_x and G_y are conjugate for every x and y in $E - G(z)$ as may be proved in a similar way to the method used to prove that G_x^* and G_y^* are conjugate.

It will therefore be sufficient to prove that $\alpha(x)$ is constant on $E - G(z)$ and it has already been observed that it is constant on certain non-vacuous subsets of $E - G(z)$.

Consider a point p^* of $E^* - z^*$ around which $\alpha^*(x^*)$ is constant. If $\alpha^*(x^*)$ is not constant everywhere then as a point moves along ray E^* in either direction it must come to a first point b^* where the function has a different value. Assume to be definite that $\alpha^*(p^*) = r$ and that for all points x^* immediately to the left of b^*,

$$\alpha^*(x^*) = r$$

whereas

$$\alpha^*(b^*) = s, \; s \neq r.$$

There must be a positive integer $t \neq 1$ such that $s = tr$, so that

$$\alpha^*(b^*) = tr, \; t \neq 1.$$

Speaking roughly this means that if x and b are points in E in the orbits corresponding to x^* and b^*, then $G(x)$ winds t times around the orbit $G(b)$. This violates intuition because these are $(n - 1)$-dimensional orbits and it will now be shown that it is in fact impossible.

Let V be that part of an open neighborhood of b which is on one side of $G(b)$, where V is so selected that for $x \in V$, $\alpha(x) = r$. For any x in V there will now be defined a transformation T_x of $G(x)$ into $G(b)$ as follows: Let g_x be an element of G such that

$$g_x G_x g_x^{-1} \subset G_b$$

and assume g_x so selected that g_x tends to e as x tends to b. The theorem in 5.3.1 on conjugate subgroups proves that such a choice is possible. Then if $y = g(x)$ is any point in $G(x)$ let

$$T_x(y) = g_x g g_x^{-1}(b).$$

This transformation is a local homeomorphism of order t taking $G(x)$ into $G(b)$.

When x is near b, T_x moves points a short distance and T_x has *degree* t for properly chosen orientations of $G(x)$ and $G(b)$ (see Alexandroff-Hopf).

The point z is inside $G(x)$ for x in V and also inside $G(b)$. These two manifolds are orientable and imbedded topologically in E. The *order* (see Alexandroff-Hopf) of z with respect to either of these manifolds has absolute value one. However the remarks above about T_x prove that the order of z with respect to one of these manifolds is t times its order with respect to the other. This is a contradiction and proves that $\alpha(x)$ is constant in $E - G(z)$ and this completes the proof of the Theorem.

THEOREM. *Let G be a compact connected group which acts on euclidean space E with at least one $(n-1)$-dimensional orbit. Then the exceptional orbit is a point z and there is a ray beginning at z which is a cross section for all orbits in the n-space E.*

We have shown that if x is near a point p of $E - G(z)$ then G_x is conjugate to G_p, and there must be a compact set A containing p which is a local cross-section of the orbits filling a neighborhood of $G(p)$. Then A must be homeomorphic to a neighborhood of p^* in E^* and since $E^* - z^*$ is a ray it follows that A can be chosen to be an arc.

Let p^*q^* and q^*r^* be arcs in $E^* - z^*$ having only q^* in common and with the arc $p^*q^*r^*$ as union. Let pq and $q_1 r_1$ be arcs in $E - G(z)$ which are cross-sections respectively of the sets of orbits represented by p^*q^* and q^*r^*. If $q_1 = q$ then pq and q_1r_1 have a union which is an arc and a cross section for the orbits of $p^*q^*r^*$. If $q_1 \neq q$, there is a g such that $g(q_1) = q$ and then

$$pq \cup g(q_1r_1)$$

is the cross-section of the orbits $p^*q^*r^*$.

Using this fact and the Heine-Borel theorem shows that for any arc in $E^* - z^*$ there is a corresponding cross-section. Therefore there is also a cross-section for the set $E^* - z^*$, and this cross-section will be denoted by Q. It still must be shown that Q can be extended to a cross-section of all orbits including $G(z)$ and that $G(z)$ is a point.

It will be shown next that $G(z)$ is a point. Assume that this is not true. Then $G(z)$ is a manifold of dimension r, $0 < r < n$, and $G(z)$ carries a *cycle* mod 2, whether or not it is orientable. Consequently the Alexander duality theorem shows that there is a cycle C in $E - G(z)$ such that C does not bound in $E - G(z)$. The cycle C is carried by some compact set X in $E - G(z)$.

Any point p in $E - G(z)$ is on the same orbit as some $q \in Q$. There is a g in G such that $g(q) = p$. Now take q as a definite point of Q and let y be a variable point of Q. The set Q can be so selected that for y in Q

$$G_y = G_q.$$

Consider the product $Q \times G(q)$ and let T map this product into $E - G(z)$ as follows

$$T : (y, g(q)) \to g(y).$$

Then T is a homeomorphism, so $E - G(z)$ is the topological product of any one orbit and a line. Consequently any compact set in $E - G(z)$ may be deformed so as to be outside any other compact set. Therefore C can be deformed in $E - G(z)$ to a position outside some sphere in E which includes $G(z)$. Therefore C cannot be *linked* with $G(z)$. This contradiction shows that $G(z)$ is a point.

It then follows that Q can be extended to a cross-section of all orbits in E simply by adjoining the point z to Q, and this completes the proof of the Theorem.

COROLLARY. *The orbits $G(x)$, $x \neq z$, have the same homology properties as an $(n - 1)$-sphere.*

This can be proved by using deformations either toward infinity or toward the point z in a manner similar to the above.

By using recent work of Borel [1] and Wang [1] it could now be shown that $G(x)$, $x \neq z$, is an $(n - 1)$-sphere and that the group G must be one of a small number of groups which can be transitive on a sphere. (See also Montgomery-Samelson [1]). It probably follows that coordinates can be introduced into E in such a way that the action of G is orthogonal. Thus the group G acting on E with at least one $(n - 1)$-dimensional orbit is topologically equivalent to a subgroup of the orthogonal group acting orthogonally, at least, if the conjecture just made is true.

6.6. Compact connected abelian groups in E_3

THEOREM. *If G is a compact connected abelian effective transformation group of a connected locally euclidean 3-space E, then G is a Lie group.*

The proof of this theorem (Montgomery-Zippin [2]) is rather long and a brief outline is given first. The group G must contain a one-parameter dense subgroup T. If G is not a Lie group it has a compact infinite zero-dimensional subgroup F. Under the action of T, E is filled with curves for which there is a local section and this local section can be chosen to be a two-cell. Furthermore the section and the group F can be selected so that F is a transformation group acting on the section. This is because F permutes the local curves generated by T. A contradiction is then reached by using the fact already established that a compact infinite zero-dimensional group can not be effective on a locally euclidean two-dimensional space (see p. 249).

As a first step the existence of a dense one-parameter subgroup T is shown.

LEMMA. *Let G be a compact connected separable metric abelian group. Then there is a one-parameter group T which is dense in G.*

Let $G = \lim G_i$ where the connecting homomorphisms are f_i

$$f_i : G_{i+1} \to G_i.$$

There is no loss of generality in assuming that the kernel of each f_i is either a finite group or a connected group. The group G_1 is a torus of some dimension and there is a one-parameter group T_1 which is dense in G_1.

If f_1 has a finite kernel then a neighborhood of the identity in G_2 is mapped isomorphically onto a neighborhood of the identity in G_1. Then a small open arc in T_1 determines a small arc in G_2 and the latter arc generates a one-parameter group T_2 which is dense in G_2.

Consider next the case where the kernel K of f_1 is connected, and is therefore a torus of some positive dimension. Then G_2 may be expressed as a direct product

$$G_2 = K \times L$$

where L is a torus group which f_1 maps onto G_1 by an isomorphism.

Let K be of dimension r and L of dimension s. There is a one-parameter group T^* in L which is mapped isomorphically onto T_1. The group T^* is everywhere dense in L. Since $L = C_1 \times C_2 \times \ldots \times C_s$, C_i a circle, any element t^* on T_1^* near e has the following form

$$t^* = (x_1, \ldots, x_s).$$

The fact that T^* is everywhere dense means that the coordinates x_i, $i = 1, \ldots, s$ are rationally independent.

If $K = C_{s+1} \times \ldots \times C_{s+r}$, it is possible to choose coordinates $x_{s+1}, \ldots, x_{s+r}$ so that x_i, $i = 1, \ldots, s+r$, are rationally independent. Then by Kronecker's theorem the one-parameter group T_2 determined by the point $x_1, \ldots, x_{s+r}$ is everywhere dense in $G_2 = L \times K$.

The group T_2 is mapped by f_1 in a one-one way on T_1.

This process may be continued and gives one-parameter groups T_i dense in G_i and satisfying $f_i(T_{i+1}) = T_i$, this relation being one-one. Hence $\lim T_i$ is a one-parameter group T dense in G and this completes the proof of the Lemma.

The action of T on E will now be considered. There must be a point p which is moved by T since G is effective.

There exists a neighborhood X of p in E and an arc J in T including e in its interor such that if x is in X and t and t' are in J and distinct then

$$t(x) \neq t'(x).$$

Hence by the theorem (of Whitney [1], see 5.4) on local cross-sections *there is a compact local cross-section K at p.*

This means that for some arc J including e in T the points $t(k)$, $t \epsilon J$, $k \epsilon K$, cover a neighborhood of p; further, this covering is unique, that is if $k' \epsilon K$, $t' \epsilon J$ and $t'(k') = t(k)$ then $t = t'$, $k = k'$.

It was proved by Whitney [2] that the set K is a two-cell in some neighborhood of p; an independent proof was given in Montgomery-Zippin [2]. The proofs call for some topological characterisation of the two-cell or of the two-sphere. The reader will find such theorems discussed fully in Wilder [1]. Here we shall sketch only briefly a way in which properties of K are inferred from properties of the containing space E. One can do this in several ways and we fall back on some local separation properties of three-space which were first studied by Alexandroff [1], and which are discussed in Wilder [1].

First, let $S \subset E$ denote a solid open sphere with p as center and which is contained in $J(K)$. Suppose further that S is small enough so S, $t_1(S)$ and $t_1^{-1}(S)$, $t_1 > 0$, are mutually exclusive. Let W denote the component of $K \cap S$ which contains p. It is possible that the set W as constructed is not simply-connected and therefore not a two-cell; but it turns out to be homeomorphic to a subset of the two-sphere.

Any subset of S can be *projected* into K, along the lines of action of J, and any subset of S which contains a point of W and is small enough will then project into W. Since S is locally connected, it follows that W is locally connected and in fact W is also locally deformable to a point. Let q denote an arbitrary point of W and consider the arc $J(q)$. Since no arc of three-space in any connected open domain of three-space can separate that domain it follows that q cannot separate any neighborhood of itself in W. Then W is cyclicly connected and, in fact, W contains arbitrarily small simple-closed curves.

If C denotes an arbitrary simple closed curve in W, then the set $J(C)$ is a compact cylinder and separates three-space locally in the neighborhood of every point which is *not an edge point*. It follows from this that there are points on W near to C which cannot belong to a connected subset of $J(W) - J(C)$. Since W is contained in S, W cannot meet $t_1(S)$ or $t_1^{-1}(S)$ and it follows without difficulty that C separates W.

On the other hand, if ab denotes an arc in W, then $J(ab)$ is a closed two-cell which does not separate space locally in the neighborhood of points of its edge and in particular it does not separate space in the neighborhood of the point a. The fact that one can go "around the edge" of the two-cell $J(ab)$ in the vicinity of the point a shows easily that one can go around the end (namely the point a) of the arc ab in the set W. The fact that this is true for an arbitrary arc ab of W shows that no arc of W can separate W in the large.

Finally, in view of known characterizations of the open subsets of the two-sphere (Zippin [1], for example; see Wilder [1], and also Whitney, loc. cit) it follows that W is a two-cell in the neighborhood of any one of its points.

It will now be shown that a small subgroup in G can be viewed as a transformation group on a properly chosen subset of W.

Let w be a point of W and let g be a small group element in G. Then

$$gJ(w) = Jg(w)$$

must intersect W in a unique point w'. Define a new action of G by

$$g\{w\} = w'.$$

Let Z be any compact zero-dimensional subgroup of G. If W^* is an open connected subset of W which includes p and if F is a compact zero-dimensional subgroup of Z so small that $F\{W^*\} \subset W$ and if for $f \in F$, $f\{W^*\}$ intersects W^* then

$$W^{**} = F\{W^*\}$$

is a connected locally euclidean two-dimensional space on which F operates as a transformation group.

It is shown in **6.4.3.** that F possesses a compact open subgroup H such that for all $x \in W^{**}$

$$H\{x\} = x.$$

This means from the definition of $g\{w\}$ that $g \in H$ implies $g\{w\}$ is in the arc $J(w)$, that is H operates on a locally euclidean one-dimensional space. Therefore

$$w \in W^{**} \text{ and } g \in H \text{ implies } g(w) = w.$$

In particular this is true when $w = p$.

But p was an arbitrary moving point. Thus if p is any moving point there is a compact open subgroup of Z which does not move p. The same is all the more true if p is not a moving point under G.

Hence for any $p \in E$, there is an open subgroup of Z which does not move p. Therefore Z is finite (see **6.3.1**).

It has been shown that G contains no infinite compact zero-dimensional subgroup. It follows that G is a Lie group. This completes the proof of the theorem.

THEOREM. *If G is a compact connected abelian group which operates effectively on euclidean three-space E, then G is a circle group.*

It is already known that G is a Lie group, and so G is the direct product of a finite number of circle groups (**2.16.4**). It will be sufficient to prove that G cannot be the product of two circle groups (assuming, as we do, that G is non-trivial).

Assume the theorem false and assume that

$$G = T_1 \times T_2$$

where T_1 and T_2 are circle groups.

For any $x \in E$ the subgroup leaving fixed all of $M = G(x)$ is the group G_x (because G is abelian). Therefore for some x, G_x is finite and the orbit $G(x)$ is two-dimensional. Such an orbit is homeomorphic to $G(x) = G/G_x$ and is a torus. This contradicts the known fact (**6.5**) that an $(n-1)$-dimensional orbit in n-space must have the homology properties of an $(n-1)$-sphere. This proves the Theorem.

This Theorem does not describe how a circle can act on E, but that will be done later and it will be shown that the action is equivalent to the group of all rotations around an axis.

6.7. Compact connected groups in E_3

THEOREM. *Let G be a compact connected group which acts on euclidean three-space E effectively. Then G is either the circle group or the proper orthogonal group in three variables.*

It should again be noted that the conclusion is only about the nature of G and says nothing about the action of G. It will be shown later that in either case the action of G is equivalent to the familiar geometric case (Montgomery-Zippin [4]).

Since G is compact and connected (assumed not trivial)

$$G = \lim G_i$$

where each G_i is a compact connected Lie group. If any G_i contains a toral group A_i then G_{i+1} contains a toral group A_{i+1} which is mapped onto A_i. This can be chosen so that A_{i+1} goes to A_i with a finite kernel (this takes argument).

Hence if any G_i contains a torus of dimension r then G contains a compact connected abelian subgroup of dimension r. By the theorem of 6.6 it follows that every G_i is of *rank* one, that is it contains no toral group of dimension greater than one.

The compact Lie groups of rank one are all well known, and it follows that each G_i is either (1) a circle group (2) the proper orthogonal group in three variables or (3) the universal covering of (2), that is the group of quaternions of absolute value one which is a double covering of (2).

If all groups G_i are of type (1) or of type (2) the theorem follows. Suppose now that some G_i is of type (3). Then each subsequent G_i is the same type and the homomorphisms f_i taking G_{i+1} to G_i are isomorphisms from this point on. Hence G itself is of type (3).

Orbits of G are either points or two-dimensional. Since G is effective some orbit is two-dimensional. By the structure of G such an orbit is covered by a two-sphere and such an orbit is either a two sphere or a projective plane. However by the theorem on $(n - 1)$-dimensional orbits in euclidean n-space it follows that the two-dimensional orbit is a two-sphere.

Let x be a point of this orbit. Then G_x is a circle. Now G has a center consisting of two elements. This center is in all G_x for x not stationary. But if x is stationary, G_x is equal to G and also contains the central group. Hence a G of type (3) cannot be effective on euclidean three-space. This proves the Theorem.

6.7.1. THEOREM. *Let G be a compact connected group (non-trivial) which acts effectively on euclidean three-space E. Then G is a circle group or the proper orthogonal group in three variables and the action of G is topologically equivalent respectively to the group of all rotations around an axis or the group of all proper orthogonal transformations.*

It is known that G itself is either (1) a circle or (2) the proper orthogonal group in three variables. It remains only to be proved that the action of these groups is equivalent to the ordinary geometric action as stated. Case (2) will be considered first. Hence G is now taken to be the proper orthogonal group acting effectively on E.

In this case an orbit of G must be either zero-dimensional or two-dimensional (a three-dimensional orbit would be a compact three-dimensional manifold imbedded in E which is impossible). Not all orbits can be zero-dimensional, that is a point, because G is effective. A two-dimensional orbit of G is either a projective plane or a two-sphere and only the latter is possible in E.

Hence there is a two-dimensional orbit $G(x)$ which is a two-sphere. Therefore G_x is a circle subgroup of G. Hence all orbits except $z = G(z)$ are two-dimensional. There is a "ray" R beginning at z which is a cross-section for all orbits and if x and y are in $E - z$ then G_x is conjugate to G_y.

Take x in $E - z$ and let A be the set of stationary points of G_x. The point z is in A and for each orbit in $E - z$ there are two points of the orbit in A. Thus A is homeomorphic to an open line and it may be assumed that one half of the line beginning at z is the ray R.

Next consider a euclidean three space E' with coordinates u, v, w. Let W denote the ray which is the non-negative w-axis. Let T be a homeomorphism taking R onto W.

The group G, being the orthogonal group in three variables can be made to act in a natural way on E'. The homeomorphism from R onto W will now be extended so that it takes all of E onto E'.

Let p be any point of E. Then p is on the orbit of a point $r \epsilon R$. There is a g in G such that $p = g(r)$. Define

$$T(g(r)) = T(p) = gT(r).$$

The homeomorphism T as thus defined shows that G in acting on E is equivalent to G acting naturally on E'.

This proves the Theorem for case (2) and case (1) will be considered next. The proof for this case is somewhat longer.

6.7.2.   Assume then that G is a circle group which acts effectively on euclidean three-space E. The proof given here follows Montgomery-Zippin [1] where a theorem of greater generality is proved.

It will be convenient to adjoin a point at infinity α to E to form S, the three-sphere. The action of G is extended to S by making α a stationary point of G

$$g(\alpha) = \alpha, \ g \epsilon G.$$

In this way G becomes a topological transformation group of S.

Let E^* and S^* be the orbit spaces of E and S and let T be the natural map from E and S to E^* and S^*.

Let F be the set of stationary points of S. Points in $S - F$ are called *moving* points. A stationary point x is said to have *period* zero, $p(x) = 0$. For any other point x, $p(x)$ is the smallest non-zero element of G which returns x to x, so that $p(x)$ can be regarded as a positive real number, $0 \leqq p(x) \leqq 1$. Of course $p(x)$ is rational, and also $p(x)$ is lower semi-continuous. The function $p(x)$ is constant on an orbit and therefore determines an associated function $p^*(x)$ defined in S^*.

If x_n tends to a moving point x and $p(x_n)$ tends to a number p, then p is an integral multiple of $p(x)$.

Let M denote the set of moving points, D the subset of M where $p(x)$ is discontinuous and C the subset of M where $p(x)$ is continuous. Then

$$S = F \cup C \cup D$$

where the three sets F, C, D are mutually disjoined. The set F is closed. The set C is open and everywhere dense in M because of the semi-continuity of $p(x)$ and the special nature of the discontinuities as mentioned above. It also follows from earlier theorems that F is nowhere dense in S.

If y is a point of C then for x near y, $G_x = G_y$ so that there is a local cross section at y from the theorem on existence of local cross-sections. Since local sections can be pieced together in the present case we have:

LEMMA 1. *If A is an arc, two-cell, or simple closed curve in $T(C)$ then $T^{-1}(A)$ has a cross-section and is the direct product of the cross-section and a simple closed curve.*

LEMMA 2. *If the arc ab is in $T(C)$ except for the point a which is in $T(F)$ then $T^{-1}(ab)$ has a cross section and is a closed two-cell.*

The proofs of these Lemmas are omitted.

6.7.3.   A type of set used below will now be described. Let B be a compact cylinder, the product of a simple closed curve and an arc. Let points on each base be specified by θ, $0 \leq \theta \leq 2\pi$. On one base of B identify each set of k points of the form $\theta + 2n\pi/k$, $n = 0, 1, \ldots, k - 1$. The set obtained in this way will be called B_k.

LEMMA. *Let ab be an arc which, except for a, is in $T(C)$ and is such that as x approaches a along ab, $p^*(x)$ approaches $kp^*(a)$. Then $T^{-1}(ab)$ contains a cross-section and is homeomorphic to B_k.*

If cd is any subarc of ab which does not include a, then $T^{-1}(cd)$ has a cross-section by Lemma 1 above (a is not fixed).

Let y be any point of $T^{-1}(a)$, and let K be a compact local cross-section in the space $T^{-1}(ab)$ of the orbits of G. This means that there is in G a symmetric arc J including e in its interior such that the points

$$J(K) = \cup t(c)\ t \epsilon J,\ c \epsilon K$$

are a product $J \times K$ and form a neighborhood of y in $T^{-1}(ab)$.

The set $T(K)$ is a compact neighborhood of a in the arc ab and $T(y) = a$. If $x \neq a$ is in ab and sufficiently near to a then $T^{-1}(x)$ cuts K in precisely k points. Choose af in ab so that for x in af, $x \neq a$, $T^{-1}(x)$ cuts K in precisely k points, and now let

$$K' = K \cap T^{-1}(af).$$

It can be seen that $K \cap T^{-1}(af - a)$ consists of exactly k half open intervals which are mutually disjoined.

Consider one of these half open intervals and add to it the point y thus obtaining a closed interval which is a cross-section of the orbits of $T^{-1}(af)$. The interval fb is such that $T^{-1}(fb)$ has a cross-section and this can be translated so as to touch the cross-section of $T^{-1}(af)$ in one point. The union of the two then is a cross-section of $T^{-1}(ab)$.

By using this cross-section it can be seen that $T^{-1}(ab)$ is homeomorphic to B_k and this proves the lemma.

6.7.4. LEMMA. *The set D in (6.7.2) is vacuous.*

This will be proved by showing that $T(D) \subset S^*$ is vacuous. Since $T(D)$ is closed in $T(M)$ it follows that

$$T(D) = D_1 \cup D_2$$

where D_1 is *perfect* in $T(M)$ and D_2 is countable. Two cases will be considered.

Case 1. D_1 is vacuous.

In this case $T(D)$ is a countable set closed in $T(M)$. Then $T(D)$ must contain an isolated point b. Since $T(F)$ is closed there exists in S^* an arc $b'b$ which except for b belongs to $T(C)$. In this case the complement of D is connected and therefore there is an arc fb' where f is in $T(F)$ and all other points are in $T(C)$. Then $fb' \cup b'b$ contains an arc fcb where f is in $T(F)$, $b \epsilon T(D)$ and all other points of fcb are in $T(C)$.

At the point b it can be seen that $p^*(x)$ has a unique limit say

$kp^*(b)$. Hence $T^{-1}(cb)$ is homeomorphic to B_k and $T^{-1}(jc)$ is a two-cell. The set

$$T^{-1}(jcb)$$

carries a non-trivial two-cycle mod k but it carries no non-trivial two-cycle with the integers as coefficients. By the Alexander duality theorem such a set cannot be imbedded in the three-sphere S. This contradiction shows that case 1 is impossible.

Case 2. D_1 is not vacuous.

The function $p^*(x)$ is semi-continuous. This function restricted to $D_1 \cup D_2$, denoted by $p^*(x_1 D_1 \cup D_2)$ is also semi-continuous. Hence the points of continuity of $p^*(x_1 D_1 \cup D_2)$ are dense in $D_1 \cup D_2$ and are a G_δ, that is the intersection of a countable number of open subsets of $D_1 \cup D_2$. Therefore there is a point b in D_1 at which $p^*(x_1 D_1 \cup D_2)$ is continuous.

Let U be an open neighborhood of b in S^* so chosen that $T(F) \cap U = \Phi$.

Let bcd be an arc in U with $b \in D_1$, d in $D_1 \cup D_2$ and $(bcd - b - d) \subset T(C)$. The function $p^*(x_1 bcd - b - d)$ is continuous and in fact constant because a function with rational values which is continuous on a connected set must be constant on the set. Hence if $x \in bcd - b - d$

$$p^*(x) = k_1 p^*(b)$$
$$p^*(x) = k_2 p^*(d).$$

Now U may be assumed as chosen in advance so that if $d \in U \cap (D_1 \cup D_2)$ then $p^*(d) = p^*(b)$. This is because $p^*(x)$ has the form $1/n$. Hence $k_1 = k_2$ and define

$$k = k_1 = k_2.$$

Then $T^{-1}(bc)$ and $T^{-1}(cd)$ are homeomorphic to B_k.

The set $T^{-1}(bcd)$ carries a non-trivial two-cycle mod k but it does not carry a non-trivial two-cycle with the integers as coefficients. As before such a set cannot be imbedded in S by the Alexander duality theorem. This shows that case 2 is impossible. This proves that D is vacuous and concludes the proof of the Lemma.

LEMMA. *The set M is connected.*

If M is not connected let

$$M = U_1 \cup U_2$$

where U_1 and U_2 are disjoined open sets, both non-vacuous. Then

$$S = F \cup U_1 \cup U_2$$

and G may be defined as a transformation group in a new way as follows

if $x \in U_1 \cup F$, let $g\{x\} = x$
if $x \in U_2$, let $g\{x\} = g(x)$.

With this definition G is a transformation group of S and since the set of stationary points includes U_1, it follows that every point is stationary. This is a contradiction and proves the Lemma because G is compact.

If $p \in F$ then there are arbitrarily small open invariant connected sets which include p.

LEMMA. *If O is an open connected set wihch is invariant under G then O — F is connected and arcwise connected.*

This can be proved in the same way as the Lemma above. It shows that F does not separate locally. Hence F is arcwise accessible from M.

LEMMA. *The set T(M) is homeomorphic to an open two-cell.*

Let J denote any simple closed curve in $T(M)$. Then $T^{-1}(J)$ is a torus and

$$S - T^{-1}(J) = U_1 \cup U_2$$

where U_1 and U_2 are disjoined, open connected sets, and

$$\overline{U}_i - U_i = T^{-1}(J), \; i = 1, \, 2.$$

Further the sets U_1 and U_2 are invariant and therefore $T(U_1)$ and $T(U_2)$ are disjoined. But

$$S^* - J = T(U_1) \cup T(U_2)$$

and therefore J separates S^* and in consequence separates $T(M)$.

Next let A be an arc in $T(M)$. Then $T^{-1}(A)$ is homeomorphic to a circle times an interval. Hence $T^{-1}(A)$ does not separate S and A does not separate S^*. Also $T^{-1}(A)$ cannot separate M and therefore A does not separate $T(M)$. Similarly it can be seen that if p is a point in $T(M)$ then p does not separate $T(M)$.

To show that $T(M)$ is a plane, use will be made of a characterization of the plane given by Zippin [1] (see also Wilder [1]). It must be shown in addition to the above that either $T(U_1)$ or $T(U_2)$ has a compact closure in $T(M)$, and that the other does not.

Either U_1 or U_2 contains the point at infinity α, say $\alpha \in U_2$. Hence $T(U_2)$ does not have a compact closure in $T(M)$.

It remains to show that U_1 does not contain points of F for then $T(U_1)$ has a compact closure in $T(M)$.

The fact that $T^{-1}(J)$ is accessible from each domain shows that the same is true for J. Assuming that $T(U_1)$ contains a point of $T(F)$ there must be an arc abc such that a and c are in $T(F)$, b is in J, the open arc ab is in $T(U_1 - F)$, and the open arc bc is in $T(U_2 - F)$. The sets $T^{-1}(ab)$, $T^{-1}(bc)$ are two-cells and their union is a two-sphere S_2 which has in common with the torus $T^{-1}(J)$ the simple closed curve $T^{-1}(b)$.

By the duality theorem S_2 separates E into exactly two open connected domains V_1 and V_2. Now $T^{-1}(J - b)$ is connected and must be entirely in one or the other of these sets, say in V_1. This implies that there cannot be points of V_2 in both U_1 and U_2. Since the points of $T^{-1}(a)$ and $T^{-1}(c)$ (points of S_2) must be accessible from V_2 this is a contradiction.

It has already been remarked that if A is a two-cell in $T(C)$ then $T^{-1}(A)$ has a cross-section (first proved by Seifert [1] and later extended by Feldbau [1]).

LEMMA. *If x is a point of $T(M)$, the basic cycle of $T^{-1}(x)$ cannot bound in M.*

Let z be the basic one-cycle of the simple closed curve $T^{-1}(x)$.

Since $T(M)$ is a plane, and since it is now known that D is vacuous, it follows that M has a cross-section in the large. This determines a homeomorphism from M onto the topological product of $(T)M$ and a circle with orbits going to circles. Hence z cannot bound in M. This shows that F is not empty.

LEMMA. *If z is a point of F then $F - z$ is connected.*

If z is a point of F for which the lemma is false then F is the union of two closed non vacuous sets X and Y (see 1.24) whose intersection is vacuous or consists only of the point z. Let x and y be in X and Y respectively and let $T(x) = a$, $T(y) = b$. Then there is an arc ab in S^* which contains no points of $T(F)$ except a and b. This follows from the fact that F is arcwise accessible.

Then $T^{-1}(ab)$ is a two-sphere and its basic two-cycle bounds in $S - z$. On the other hand if a_1 is an interior point of ab, the basic cycle of $T^{-1}(a_1)$ bounds in the complement of X and in the complement of Y. This however contradicts Lemma W^i of a paper by Alexander [2], and this contradiction establishes the Lemma.

It follows from the previous Lemma that F contains more than two points, and now it follows that F is connected.

LEMMA. *The set $T(E)$ is homeomorphic to a closed half plane.*

Take x and y in F as in the preceding Lemma. Again let $T(x) = a$, $T(y) = b$, and let ab be an arc with no points in $T(F)$ except a and b. Then $T^{-1}(ab)$ separates S so that the arc ab separates $S^* = T(S)$. A proper subset of the arc corresponds to a proper subset of the two-sphere and does not separate. Hence ab separates *irreducibly.*

It is true that each domain of $S - T^{-1}(ab)$ contains a simple closed curve orbit. Since this orbit bounds in the domain, it follows that each of the domains of $S - T^{-1}(ab)$ contains a point of F. Therefore x and y separate F.

Therefore $T(F)$ is a continuum without cut points which is separated by every pair of points. Hence (Whyburn [1]) $T(F)$ and F are simple closed curves. It follows from a theorem of Zippin (see Wilder [1]) that $T(S) = S^*$ is a closed two-cell whose boundary is $T(F)$. The fact that $T(E)$ is a closed half plane with boundary $T(F - \alpha)$ follows and this proves the Lemma.

The argument for the theorem can soon be concluded.

It has already been remarked that M has a cross-section which will now be denoted by K_1. Then

$$T : K_1 \to T(M)$$

is a homeomorphism of K_1 onto $T(M)$. However T is also a homeomorphism of $F - \alpha$ onto $T(F - \alpha)$.

Let

$$K = K_1 \cup (F - \alpha).$$

Then T maps K onto $T(E)$ and it can be seen that the map is a homeomorphism.

Any point x of E is on the orbit of a point k in K, that is, there is $g \in G$ such that

$$g(k) = x.$$

The group G also acts in a natural way on three-space y_1, y_2, y_3 as a rotation around the y_3-axis. In this natural action each point is on the orbit of a point in the half plane P given as follows

$$P : y_1 = 0, \ y_2 \geq 0, \ y_3 \text{ arbitrary.}$$

There is a homeomorphism T^* from K to the half-plane P taking $F - \alpha$ to the axis $y_1 = 0$, $y_2 = 0$. It may now be left to the reader to extend T^* to be a homeomorphism from all of E onto all of the space y_1, y_2, y_3, and in such a way that the original action of G is carried to the action on (y_1, y_2, y_3). This proves the Theorem.

6.8. One-parameter transformation groups in E_3

As already remarked, the action of V_1 on the plane or on three-space can be very involved, and extra conditions must be imposed in order to achieve useable theorems. Two results of this kind will be described without proof.

A one-parameter transformation group V_1 of E_3 is called *dispersive* if for every pair of points x, $y \in E_3$ there exist neighborhoods U_x, $U_y \subset E_3$ and a positive number N such that for every $z \in U_x$ and every $v \in V_1$, $|v| > N$, the point vz is *not* in U_y (Montgomery-Zippin [3]).

THEOREM 1. *A necessary and sufficient condition that V_1 be a translation-group of E_3 (in an appropriate coordinate-system) is that it be dispersive.*

The second Theorem rests on a definition given in 1.26.3, example 6, of Chapter I.

THEOREM 2. *A one-parameter transformation group of E_3 which is pointwise periodic, whose orbits are small near fixed points and whose period function is bounded in the neighborhood of moving points, is a topological quasi-rotation group of E_3.*

The proof of the second theorem does not differ very much from the analysis of the circle group on E_3 (Montgomery-Zippin [1]). In the first Theorem the key idea of the proof is to construct a cross-section in the large.

6.9. Rigid geometries of E_3

When a compact group acts on a metric space it is easy to find a metric for the space which is invariant under the action of the group. Thus, let $d(x, y)$ be a metric for a space M acted on by a compact group G. Choose a left-invariant integral in G (2.17) normalised by the condition $\int 1 = 1$. Now define a new metric $d*$ by

$$d*(x, y) = \int d(gx, gy).$$

Then from the triangle inequality for d expressed by

$$d(gx, gy) + d(gy, gz) \geqq d(gx, gz)$$

for every $x, y, z \in M$ and $g \in G$, the triangle inequality for $d*$ follows at once. The other properties of a metric are equally obvious. The invariance of $d*$, for every $h \in G$, follows from

$$d*(hx, hy) = \int d(ghx, ghy) = \int d(g'x, g'y) = d*(x, y).$$

Of course, if the metric d has certain special properties, it does not follow that $d*$ will have them (in the Bing example in E_3 already referred to, the "averaging process" just described *must* alter the given euclidean metric to something that cannot be euclidean).

We may express what has been proved above by saying that every compact group of transformations of a metric space M can be regarded as a subgroup of a group of isometries of M. This is not true in general for locally compact groups, and not even for the projective group on a line, since that group has one-parameter subgroups which cause a *moving* point on the line to approach a *fixed* point.

6.9.1.   The notion of *congruence* which underlies the euclidean and hyperbolic geometries of the plane and of three-space is essentially a concept of isometry applied to transformation groups which are not given as topological groups (but which acquire a natural topology from the way in which they operate on a topological space). The topology which is appropriate to these transformation groups makes them locally compact.

In one of his studies of the geometry of the plane Hilbert [2] formulates a class of transformation groups which may be described as follows. Let M be a topological space and let G be a group of homeomorphisms of M. A pair $x, y \in M$ is called *congruent* to a pair $x', y' \in M$ if there is a $g \in G$ such that $gx = x'$ and $gy = y'$. A pair $x, y \in M$ is called *semicongruent* to a pair $x'', y'' \in M$ if with respective neighborhoods $U(x)$, $V(y)$, $U(x'')$, $V(y'')$, all in M, there exists a, b in $U(x)$, $V(y)$ and a'', b'' in $U(x'')$, $V(y'')$ such that the pair (a, b) is congruent to the pair a'', b''.

Hilbert studied transformation groups of the number-plane which had the property that every semicongruent pair (of pairs of points) was also congruent. His work was partly carried over to three-space by Kerekjarto [2] in 1927. After the structure of compact topological groups had been worked out it was possible to extend Hilbert's theorem to three-space in a form very close to the original (Montgomery-Zippin [7]) as follows.

THEOREM.   *Suppose that E is E_3 and that G is a group of homeomorphisms of E onto itself such that*

1) *if a pair of pairs of points of E is semicongruent under G then it is congruent*

2) *there is at least one point $p \in E$ such that $G_p \subset G$ is a proper subgroup of G, and for this point*

2a) *the set $G_p(x)$ is infinite, $x \neq p$*

2b) *there is at least one $q \in E$, $q \neq p$, such that $G_p(q)$ is not totally disconnected.*

CONCLUSION: *line, plane, angle, congruence may all be defined in E and E is the euclidean or the hyperbolic space and G is the group of all congruences or it is the group of all orientation preserving congruences.*

We have remarked before that it is not known whether a compact totally disconnected infinite group can operate effectively on three-space. In line with this, it is not clear whether condition 2b) is redundant. However, in the present theorem we are dealing with a rather large group of transformations of the three-space and the question may be a little easier to answer.

The condition 2b) *is* superfluous in the plane (6.4.2) and Hilbert does not use it in his theorem. The remaining conditions, above, are a little weaker than the ones used by Hilbert (for the plane) but they are sufficient to give his result if E, above, is interpreted to be the number-plane (rather than E_3).

Concerning the proof we shall only remark that the group G_p is first proved to be compact (in a naturally defined topology which embraces all of G). Then the preceding results give the structure of the group G_p and of each G_x for which $x \in E$ can be proved to be on the orbit $G(p)$. Ultimately, of course, G is seen to be transitive on the space.

BIBLIOGRAPHY

ADO, I.
[1] On the representation of finite continuous groups by means of linear transformations. Izvestia Kazan 7 (1934—35), pp. 3—43 (in Russian).

ALEXANDER, J. W.
[1] On the characters of discrete abelian groups. Ann. of Math. (2) 35 (1934), pp. 389—395.
[2] A proof and extension of the Jordan-Brouwer separation theorem. Trans. Amer. Math. Soc. 23 (1922), pp. 333—349.

ALEXANDER, J. W., and ZIPPIN, L.
[1] Discrete abelian groups and their character groups. Ann. of Math. (2) 36 (1935), pp. 71—85.

ALEXANDROFF, P.
[1] On local properties of closed sets. Ann. of Math. (2) 36 (1935), pp. 1—35.

ALEXANDROFF, P., and HOPF, H.
[1] Topologie I, Berlin, 1935.

ARENS, R.
[1] Topologies for homeomorphism groups. Amer. J. Math. 68 (1946), pp. 593—610.

BING, R.
[1] A homeomorphism between the 3-sphere and the sum of two solid horned spheres. Ann. of Math. 56 (1952), pp. 354—362.

BIRKHOFF, G.
[1] Analytical groups. Trans. Amer. Math. Soc. 43 (1938), pp. 61—101.
[2] Lie groups simply isomorphic with no linear group. Bull. Amer. Math. Soc. 42 (1936), pp. 883—888.
[3] A note on topological groups. Compositio Math. 3 (1936), pp. 427—430.

BOCHNER, S.
[1] Compact groups of differentiable transformations. Ann. of Math. (2) 46 (1945), pp. 372—381.
[2] Formal Lie groups. Ann. of Math. (2) 47 (1946), pp. 192—201.

BOCHNER, S., and MONTGOMERY, D.
[1] Groups of differentiable and real or complex analytic transformations. Ann. of Math. 46 (1945), pp. 685—694.
[2] Locally compact groups of differentiable transformations. Ann. of Math. 47 (1946), pp. 639—653.
[3] Groups on analytic manifolds. Ann. of Math. (2) 48 (1947), pp. 659—669.

BOREL, A.
[1] Some remarks about transformation groups transitive on spheres and tori. Bull. Amer. Math. Soc. 55 (1949), pp. 580—586.
[2] Les bouts des espaces homogènes de groupes de Lie, Ann. of Math. 58 (1953), pp. 443—457.
[3] Sur la cohomologie des espaces fibrés principaux et des espaces homogènes de groupes de Lie compacts. Ann. of Math. 57 (1953), pp. 115—207.

BOREL, A., and DE SIEBENTHAL, J.
[1] Les sousgroupes fermés connexes de rang maximum des groupes de Lie clos. Comment. Math. Helv. 23 (1949—50), pp. 200—221.

BORSUK, K.
[1] Über die Fundamentalgruppe der Polyeder. Monatsh. Math. Phys. 41 (1936), pp. 64—77.

BOURBAKI, N.
[1] Topologie générale. Paris, 1942.

BROUWER, L. E. J.
[1] Die Theorie der endlichen kontinuierlichen Gruppen unabhängig von den Axiomen von Lie. (2 papers.) Math. Ann. 67 (1909), pp. 246—267; 69 (1910), pp. 181—203.
[2] Über die periodischen Transformationen der Kugel. Math. Ann. 80 (1921), pp. 39—41.
[3] On the structure of perfect sets of points. Proc. Acad. Amsterdam 12, pt. 2 (1910), pp. 785—794.

BUSEMAN, H.
[1] Metric methods in Finsler spaces and in the foundations of geometry. Princeton, Univ. Press, 1942.

CARTAN, E.
[1] Oeuvres Completes. Part 1, v. 1, 2, Paris 1952.
[2] La topologie des groupes de Lie. Hermann, Paris, 1936.

CARTAN, H.
[1] Sur les groupes de transformations analytiques. Hermann, Paris, 1935.

CHEVALLEY, C.
[1] Theory of Lie groups. Princeton Univ. Press, 1946.
[2] Two theorems on solvable topological groups. Michigan Lectures in Topology (1941), pp. 291—292.
[3] On a theorem of Gleason. Proc. Amer. Math. Soc. 2 (1951), pp. 122—125.

CHEVALLEY, C., and EILENBERG, S.
[1] Cohomology theory of Lie groups and Lie algebras. Trans. Amer. Math. Soc. 63 (1948), pp. 85—124.

VAN DANTZIG, D.
[1] Zur topologischen Algebra, I. Math. Ann. 107 (1933), pp. 587—626.
[2] Über topologisch homogene Kontinua. Fund. Math. 15 (1930), pp. 102—125.

VAN DANTZIG, D., and VAN DER WAERDEN.
[1] Über metrische homogene Räume. Abh. Math. Sem. Hamb. Univ. 6 (1928), pp. 367—376.

EHRESMANN, C.
[1] Sur les espaces localement homogènes. L'Enseignment Math. 35 (1936), pp. 317—333.
[2] Introduction à la théorie des structures infinitesimales et des pseudogroupes de Lie. Colloque Intern. de Géom. Differentielle de Strasbourg (1953).
[3] Sur la topologie de certain espaces homogènes. Ann. of Math. 35 (1934), pp. 396—443.

EILENBERG, S.

[1] Sur les transformations periodiques de la surface de sphere. Fund. Math. 22 (1934), pp. 28—41.

[2] On a theorem of P. A. Smith concerning fixed points for periodic transformations. Duke Math. J. 6 (1940), pp. 428—437.

EILENBERG, S., and MacLANE, S.

[1] Group extensions and homology. Ann. of Math. 43 (1942), pp. 757—831.

EISENHART, L. P.

[1] Continuous groups of transformations. Princeton, Univ. Press, 1933.

FELDBAU, J.

[1] Sur la classification des espaces fibrés. C. R. Acad. Sci. Paris 208 (1939), pp. 1621—1623.

FLOYD, E. E.

[1] Examples of fixed point sets of periodic maps. Ann. Math. 55 (1952), pp. 167—171.

[2] On related periodic maps. Amer. J. Math. 74 (1952), pp. 547—554.

FREUDENTHAL, H.

[1] Einige Sätze über topologische Gruppen. Ann. of Math. (2) 37 (1936), pp. 46—56.

[2] Entwicklungen von Räumen und ihren Gruppen. Compositio Math. 4 (1937), pp. 145—234.

[3] La structure des groupes à deux bouts et des groupes triplement transitifs. Proc. Acad. Amst. 54 (1951), pp. 288—294.

GLEASON, A.

[1] Arcs in locally compact groups. Proc. Nat. Acad. Sci. U.S.A. 36 (1950), pp. 663—667.

[2] Spaces with a compact Lie group of transformations. Proc. Amer. Math. Soc. 1 (1950), pp. 35—43.

[3] Square roots in locally euclidean groups. Bull. Amer. Math. Soc. 55 (1949), pp. 446—449.

[4] On the structure of locally compact groups. Duke Math. J. 18 (1951), pp. 85—104.

[5] Groups without small subgroups. Ann. of Math. 56 (1952), pp. 193—212.

GOTO, M.

[1] On local Lie groups in a locally compact group. Ann. of Math. 54 (1951), pp. 94—95.

[2] Faithful representations of Lie groups, II. Nagoya Math. J. 1 (1950), pp. 91—101.

GOTO, M., and YAMABE, H.

[1] On continuous isomorphisms of topological groups. Nagoya Math. J. 1 (1950), pp. 109—111.

GOTTSCHALK, W. H. and HEDLUND, G. A.

Topological Dynamics. Amer. Math. Sci. Coll. Publication.

HAAR, A.

[1] Der Massbegriff in der Theorie der kontinuierlichen Gruppen. Ann. of Math. (2) 34 (1933), pp. 147—169.

HILBERT, D.

[1] Mathematische Probleme, Nachr. Akad. Wiss. Göttingen 1900, pp. 253—297.

[2] Grundlagen der Geometrie. 7th ed. (1930), Anhang IV.

HOPF, H.

[1] Über die Topologie der Gruppenmannigfaltigkeiten und ihre Verallgemeinerungen. Ann. of Math. 42 (1941), pp. 22—52.

HOPF, H., and SAMELSON, H.

[1] Ein Satz über die Wirkungsräume geschlossener Liescher Gruppen, Comment. Math. Helv. 13 (1940, 1941), pp. 240—251.

HUREWICZ, W., and WALLMAN, H.

[1] Dimension Theory. Princeton, Univ. Press 1941.

IWASAWA, K.

[1] On some types of topological groups. Ann. of Math. (2) 50 (1949), pp. 507—557.

[2] Topological groups with invariant compact neighborhoods of the identity. Ann. of Math. 54 (1951), pp. 345—348.

KAKUTANI, S.

[1] Über die Metrization der topologischen Gruppen. Proc. Imp. Acad. Jap. 12 (1936), pp. 82—84.

KAKUTANI, S., and KODAIRA, K.

[1] Über das Haarsche Mass in der lokal bikompakten Gruppen. Proc. Imp. Acad. Tokyo 20 (1944), pp. 444—450.

VAN KAMPEN, E. R.

[1] Locally bicompact abelian groups and their character groups. Ann. of Math. 36 (1935), pp. 448—463.

[2] The structure of a compact connectial group. Amer. J. Math. 57 (1935), pp. 301—308.

KAPLAN, S.

[1] A zero-dimensional topological group with a one-dimensional factor-group. Bull. Amer. Math. Soc. 54 (1948), pp. 964—968.

KAPLAN, W.

[1] Differentiability of regular curve families on the sphere. Duke Math. J. 7 (1940), pp. 154—185.

KAPLANSKY, I.

[1] Infinite Abelian Groups. University of Michigan, 1954.

KEREKJARTO, B.

[1] Über die periodischen Transformationen der Kreisscheibe und der Kugelfläche. Math. Ann. 80 (1921), pp. 36—38.

[2] On a geometrical theory of continuous groups, II: euclidean and hyperbolic groups of three-dimensional space. Ann. of Math. (2) 29 (1927), pp. 169—179.

[3] Geometrische Theorie der Zweigliedrigen kontinuierlichen Gruppen. Hamburg Abh. 8 (1931), pp. 107—114.

[4] Sur l'existence de racines carrées dans les groupes continus. C. R. Acad. Sci. Paris 193 (1931), pp. 1384—1385.

[5] Sur la structure des transformations topologiques des surfaces en elles-memes. L'Enseignement Math. 35 (1936), pp. 297—316.

KURANISHI, M.
[1] On local euclidean groups satisfying certain conditions. Proc. Amer. Math. Soc. 1 (1950), pp. 372—380.
[2] On conditions of differentiability of locally compact groups. Nagoya Math. J. 1 (1950), pp. 71—81.

LEFSCHETZ, S.
[1] Algebraic Topology. Amer. Math. Soc. Coll. Publication 27 (1942).

LEJA, F.
[1] Sur la notion de groupe abstrait topologique. Fund. Math. 9 (1927), pp. 37—44.

LIAO, S. D.
[1] A theorem on periodic transformations of homology spheres, Ann. of Math. 56 (1952), pp. 68—83.

MALCEV, A.
[1] On solvable topological groups. Mat. Sbornik (Recueil Math.) N.S. (1946), pp. 165—174.

MARKOFF, A.
[1] Über endlich-dimensionale Vektor Räume. Ann. of Math (2) 36 (1935), pp. 464—506.

MATSUSHIMA, Y.
[1[On the faithful representations of Lie groups. J. Math. Soc. Jap. 1 (1949), pp. 254—261.

MONTGOMERY, D.
[1] Pointwise periodic homeomorphisms. Amer. J. Math. 59 (1937), pp. 118—120.
[2] Topological groups of differentiable transformations. Ann. of Math. 46 (1945), pp. 382—387.
[3] A theorem on locally euclidean groups. Ann. of Math. 48 (1947), pp. 650—659.
[4] Connected one-dimensional groups. Ann. of Math. 49 (1948), pp. 110—000.
[5] Analytic parameters in three-dimensional groups. Ann. of Math. 49 (1948), pp. 118—131.
[6] Subgroups of locally compact groups. Amer. J. Math. 70 (1948), pp. 327—332.
[7] Dimensions of factor-spaces. Ann. of Math. 49 (1948), pp. 373—378.
[8] Theorems on the topological structure of locally compact groups. Ann. of Math. 50 (1949), pp. 570—580.
[9] Connected two-dimensional groups. Ann. of Math. 52 (1950), pp. 591—605.
[10] Finite-dimensional groups. Ann. of Math. 52 (1950), pp. 591—605.
[11] Locally homogeneous spaces. Ann. of Math. 52 (1950), pp. 261—271.
[12] Simply connected homogeneous spaces. Proc. Amer. Math. Soc. 1 (1950), pp. 467—469.

MONTGOMERY, D., and SAMELSON, H.
[1] Transformation groups of spheres. Ann. of Math. 44 (1943), pp. 454—470.
[2] Groups transitive on the n-dimensional torus. Bull. Am. Math. Soc. 49 (1943), pp. 455—456.

MONTGOMERY, D., and ZIPPIN, L.

[1] Periodic one-parameter groups in three-space. Trans. Amer. Math. Soc. 40 (1936), pp. 24—36.

[2] Compact abelian transformation groups. Duke Math. J. 4 (1936), pp. 363—373.

[3] Translation groups in three-space. Amer. J. Math. 49 (1937), pp. 121—128.

[4] Non-abelian, compact, connected transformation groups of three-space. Amer. J. Math. 61 (1939), pp. 375—387.

[5] Theorem on the rotation group. Bull. Amer. Math. Soc. 45 (1940), pp. 520—521.

[6] Topological transformation groups. Ann. of Math. 41 (1940), pp. 778—791.

[7] Topological group foundations of rigid space geometry. Trans. Amer. Math. Soc. 48 (1940), pp. 21—49.

[8] Theorem on Lie groups. Bull. Amer. Math. Soc. 48 (1942), pp. 448—452.

[9] A class of transformation groups. Amer. J. Math. 65 (1943), pp. 601—608.

[10] Existence of subgroups isomorphic to the real numbers. Ann. of Math. 53 (1951), pp. 298—326.

[11] Two-dimensional subgroups. Proc. Amer. Math. Soc. 2 (1951), pp. 822—838.

[12] Four-dimensional groups. Ann. of Math. 56 (1952), pp. 140—166.

[13] Small subgroups of finite-dimensional groups. Ann. of Math. 56 (1952), pp. 213—241.

[14] Examples of Transformation groups, Proc. Amer. Math. Soc. 5 (1954), pp. 460—465.

MOSTOW, G. D.

[1] The extensibility of local Lie groups of transformations and groups on surfaces. Ann. of Math. 52 (1950), pp. 606—636.

[2] A new proof of E. Cartan's theorem on the topology of semi-simple groups. Bull. Amer. Math. Soc. 55 (1949), pp. 969—980.

MYERS, S. B., and STEENROD, N. E.

[1] The group of isometries of a Riemannian manifold. Ann. of Math. 40 (1939), pp. 400—416.

NEWMAN, M. H. A.

[1] A theorem on periodic transformations of spaces, Quart. J. Math. 2 (1931), pp. 1—8.

PETER, F., and WEYL, H.

[1] Die Vollständigkeit der primitiven Darstellungen einer geschlossen kontinuierlichen Gruppe. Math. Ann. 97 (1927), pp. 737—755.

PONTRJAGIN, L.

[1] Topological groups. Princeton, Univ. Press 1939.

[2] Über die topologische Struktur der Lieschen Gruppen. Comment. Math. Helv. 13 (1940—1941), pp. 227—238.

SAMELSON, H.

[1] A note on Lie groups. Bull. Amer. Math. Soc. 52 (1946), pp. 870—873.

[2] Topology of Lie groups. Bull. Amer. Math. Soc. 58 (1952), pp. 2—37.

[3] Beiträge zur Topologie der Gruppenmannigfaltigkeiten. Ann. of Math. 42 (1941), pp. 1091—1137.

SCHREIER, O.

[1] Abstrakte kontinuierliche Gruppen. Hamburg Abh. 4 (1925), pp. 15—32.

SEGAL, I. E.

[1] The group algebra of a locally compact group. Trans. Amer. Math. Soc. 61 (1947), pp. 69—105.

[2] Topological groups in which multiplication on one side is differentiable. Bull. Amer. Math. Soc. 52 (1946), pp. 481—487.

SEIFERT, H.

[1] Topologie dreidimensionaler gefaserter Räume. Acta Math. 60 (1932), pp. 147—238.

SMITH, P. A.

[1] Fixed point theorems for periodic transformations, Amer. J. Math. 63 (1941), pp. 1—8.

[2] Transformations of finite period, III, Newman's theorem. Ann. of Math. 42 (1941), pp. 446—458.

[3] The topology of transformation groups. Bull. Amer. Math. Soc. 44 (1938), pp. 497—514.

[4] Foundations of the theory of Lie groups with real parameter. Ann. of Math. 44 (1943), pp. 481—513.

[5] Periodic and nearly periodic transformations. Michigan Lectures in Topology (1941), pp. 159—190.

SMITH, P. A., and RICHARDSON, M.

[1] Periodic transformations of complexes. Ann. of Math. 39 (1938), pp. 611—633.

STEENROD, N. E.

[1] The Topology of Fibre Bundles. Princeton Univ. Press 1951.

TYCHONOFF.

[1] Über einen Funktionenraum. Math. Ann. III (1935), pp. 762—766.

VON NEUMANN, J.

[1] Über die analytischen Eigenschaften von Gruppen linearen Transformationen und ihrer Darstellungen. Math. Zeit. 30 (1929), pp. 3—42.

[2] Die Einführung analytischer Parameter in topologischen Gruppen. Ann. of Math. 34 (1933), pp. 170—190.

[3] Zum Haarschen Mass in topologischen Gruppen. Compositio Math. 1 (1934), pp. 106—114.

[4] The uniqueness of Haar's measure. Mat. Sbornik 1 (1936), pp. 721—734.

WANG, H. C.

[1] Two-point homogeneous spaces. Ann. of Math. 55 (1952), pp. 177—191.

[2] Homogeneous spaces with non-vanishing Euler characteristics. Ann. of Math. 50 (1944), pp. 925—953.

WEIL, A.

[1] L'integration dans les groupes topologiques et ses applications. Paris, 1940.

[2] Sur les espaces à structure uniforme et sur la topologie générale. Hermann, Paris, 1938.

WEYL, H.

[1] Theorie der Darstellung kontinuierlichen halbeinfacher Gruppen durch lineare Transformationen, I. Math. Zeit. 23 (1925), pp. 271—309.

[2] Continuous Groups. Vols. 1 and 2, Institute for Advanced Study Notes, 1934.
WHITEHEAD, J. H. C.
[1] Convex regions in the geometry of paths. Quart. J. Math. 3 (1932), pp. 33—42.
WHITNEY, H.
[1] Regular families of curves. Ann. of Math. 34 (1933), pp. 244—270.
[2] Cross-sections of curves in 3-space. Duke Math. J. 4 (1939), pp. 222—226.
WHYBURN, G. T.
[1] Analytic Topology, Amer. Math. Soc. Coll. Publication 28 (1942).
[2] On sequences and limiting sets, Fund. Math. 25 (1935), pp. 408—426.
WILDER, R. L.
[1] Topology of Manifolds, Amer. Math. Soc. Coll. Publication 32 (1949).
[2] On free subsets of E_n. Fund. Math. 25 (1935), pp. 200—208.

YAMABE, H.
[1] On an arcwise connected subgroup of a Lie group. Osaka Math. J. 2 (1950), pp. 13—14.
[2] Note on locally compact groups. Osaka Math. J. 3 (1951), pp. 77—82.
[3] Generalization of a theorem of Gleason. Ann. of Math. 58 (1953), pp. 351—365.
[4] On the conjecture of Iwasawa and Gleason. Ann. of Math. 58 (1953), pp. 48—54.

ZIPPIN, L.
[1] Continuous curves and the Jordan curve theorem. Amer. J. Math. 52 (1930), pp. 331—350.
[2] Transformation groups. Michigan Lectures in Topology, 1941, pp. 191—221.
[3] Two-ended topological groups. Proc. Amer. Math. Soc. 1 (1950), pp. 309—315.
[4] Countable torsion groups. Ann. of Math. 36 (1935), pp. 86—99.

SUPPLEMENTARY BIBLIOGRAPHY
Second Printing, February, 1964

ANDERSON, R. D.

On homeomorphisms as products of conjugates of a given homeomorphism and its inverse. Topology of 3-manifolds and related topics. Proc. of The University of Georgia Institute (1961), 231—234. Prentice-Hall, Inc., Englewood Cliffs, N. J., 1962.

AUSLANDER, LOUIS

Solvable Lie groups acting on nilmanifolds. Amer. J. Math. (4) 82 (1960), 653—660.

Bieberbach's theorem on space groups and discrete uniform subgroups of Lie groups, II. Amer. J. Math. (2) 83 (1961), 276–280.

AUSLANDER, LOUIS; GREEN, L.; and HAHN, F.

Flows on some three dimensional homogeneous spaces. Bull. Amer. Math. Soc. (5) 67 (1961), 494—497.

BING, R. H.

Decompositions of E^3 into points and tame arcs. Summer Inst. on set theoretic topology, Univ. of Wisc. (1955), 41—48.

The cartesian product of a certain non-manifold and a line is E^4. Bull. Amer. Math. Soc. 64 (1958), 82—84.

BOTT, RAOUL

The space of loops on a Lie group. Mich. Math. J. 5 (1958), 35—61.

The stable homotopy of the classical groups. Ann. of Math. (2) 70 (1959), 313—337.

Quelques remarques sur les théorèmes de périodicité. Bull. Soc. Math. France 87 (1959), 293—310.

BOTT, RAOUL and SAMELSON, HANS

Applications of Morse theory to symmetric spaces. Symposium internacional de topología algebraica [International symposium on algebraic topology], pp. 282—284. Universidad Nacional Autónoma de México and UNESCO, Mexico City, 1958.

Applications of the theory of Morse to symmetric spaces. Amer. J. Math. (4) 80 (1958), 964–1029.

Correction to "Applications of the theory of Morse to symmetric spaces" (This Journal, vol. 80 (1958), 964—1029). Amer. J. Math. (1) 83 (1961), 207—208.

BOREL, ARMAND

Topology of Lie groups and characteristic classes. Bull. Amer. Math. Soc. 61 (1955), 397—432.

Nouvelle démonstration d'un théorème de P. A. Smith. Comment. Math. Helv. 29 (1955), 27—39.

Transformation groups with two classes of orbits. Proc. Nat. Acad. Sci. U.S.A. 43 (1957), 983—985.

Fixed points of elementary commutative groups. Bull. Amer. Math. Soc. 65 (1959), 322—326.

Commutative subgroups and torsion in compact Lie groups. Bull. Amer. Math. Soc. 66 (1960), 285—288.

[280]

BOREL, ARMAND and HIRZEBRUCH, F.

Characteristic classes and homogeneous spaces. I. Amer. J. Math. 80 (1958), 458—538.

Characteristic classes and homogeneous spaces. II. Amer. J. Math. 81 (1959), 315—382.

Characteristic classes and homogeneous spaces. III. Amer. J. Math. 82 (1960), 491—504.

BREDON, GLEN E.

Some theorems on transformation groups. Ann. of Math. (2) 67 (1958), 104—118.

Orientation in generalized manifolds and applications to the theory of transformation groups. Mich. Math. J. 7 (1960), 35—64.

On homogeneous cohomology spheres. Ann. of Math. (2) 73 (1961), 556—565.

Cohomology fibre spaces, the Smith-Gysin sequence and orientation in generalized manifolds. Mich. Math. J. 10 (1963), 321—333.

Transformation groups on spheres with two types of orbits. To appear.

Examples of differentiable group actions. To appear.

BREDON, GLEN E.; RAYMOND, F.; and WILLIAMS, R. F.

p-adic groups of transformations. Trans. Amer. Math. Soc. (3) 99 (1961), 488—498.

CONNELL, E. H.; MONTGOMERY, D.; and YANG, C. T.

Compact groups in E^n. To appear in Ann. of Math.

CONNER, P. E.

Concerning the action of a finite group. Proc. Nat. Acad. Sci. U.S.A. 42 (1956), 349—351.

On the action of a finite group on $S^n \times S^n$. Ann. of Math. (2) 66 (1957), 586—588.

On the action of the circle group. Mich. Math. J. 4 (1957), 241—247.

On a theorem of Montgomery and Samelson. Proc. Amer. Math. Soc. 9 (1958), 464—466.

Orbit spaces of circle groups of transformations. Ann. of Math. (2) 67 (1958), 90—98.

Transformation groups on a $K(\pi, 1)$. II. Mich. Math. J. 6 (1959), 413—417.

Orbits of uniform dimension. Mich. Math. J. 6 (1959), 25—32.

Retraction properties of the orbit space of a compact topological transformation group. Duke Math. J. (3) 27 (1960), 341—358.

Diffeomorphisms of period two. Mich. Math. J. 10 (1963), 341—352.

A bordism theory for actions of an abelian group. Bull. Amer. Math. Soc. (2) 69 (1963), 244—247.

Pontrjagin numbers of maps. Bull. Amer. Math. Soc. (2) 69 (1963), 276—279.

CONNER, P. E. and FLOYD, E. E.

A characterization of generalized manifolds. Mich. Math. J. 6 (1959), 33—43.

On the construction of periodic maps without fixed points. Proc. Amer. Math. Soc. 10 (1959), 354—360.

Differentiable periodic maps. Bull. Amer. Math. Soc. 68 (1962), 76—86.

Differentiable Periodic Maps. To appear in Ergebnisse-der-Mathematik Series, Vol. 33, Springer.

CONNER, P. E. and MONTGOMERY, D.

Transformation groups on a $K(\pi, 1)$. I. Mich. Math. J. 6 (1959), 405—412.

An example for $SO(3)$. Proc. Nat. Acad. Sci. (11) 48 (1962), 1918—1922.

DE GROOT, J.
The action of a locally compact group on a metric space. Nieuw Archief voor Wiskunde (3) 7 (1959), 70—74.

EDWARDS, C. H., JR.
Products of pseudo cells. Bull. Amer. Math. Soc. (6) 68 (1962), 583—584.
EELLS, JAMES and KUIPER, N.
An invariant for certain smooth manifolds. Annali di Matematica, Serie IV, 60 (1963), 93—110.
ELLIS, ROBERT
Continuity and homeomorphism groups. Proc. Amer. Math. Soc. 4 (1953), 969—973.
A note on the continuity of the inverse. Proc. Amer. Math. Soc. 8 (1957), 372—373.

FLOYD, E. E.
On periodic maps and the Euler characteristics of associated spaces. Trans. Amer. Math. Soc. 72 (1952), 138—147.
Orbit spaces of finite transformation groups. I. Duke Math. J. 20 (1953), 563–567.
Orbit spaces of finite transformation groups. II. Duke Math. J. 22 (1955), 33—38.
Examples of fixed point sets of periodic maps. II. Ann. of Math. (2) 64 (1956), 396—398.
Orbits of torus groups operating on manifolds. Ann. of Math. (2) 65 (1957), 505—512.
Fixed point sets of compact abelian Lie groups of transformations. Ann. of Math. (2) 66 (1957), 30—35.
Closed coverings in Čech homology theory. Trans. Amer. Math. Soc. 84 (1957), 319—337.
Some connections between cobordism and transformation groups. Proc. International Congress of Mathematicians (1962), 462—466.
FLOYD, E. E. and RICHARDSON, R. W., JR.
An action of a finite group on an n-cell without stationary points. Bull. Amer. Math. Soc. 65 (1959), 73—76.
Fox, R. H.
Knots and periodic transformations. Ibid., 177—182.

HELGASON, SIGURDUR
Differential Geometry and Symmetric Spaces. Academic Press, New York and London, 1962.
HIRSCH, MORRIS W. and MILNOR, JOHN W.
Some curious involutions of spheres. To appear.
HIRSCH, MORRIS W. and SMALE, STEPHEN
On involutions of the 3-sphere. Amer. J. Math. 81 (1959), 893—900.
HOMMA, T. and KINOSHITA, S.
On homeomorphisms which are regular except for a finite number of points. Osaka Math. Jour. 7 (1955), 29—38.
HSIANG, W. Y.
Classification of action of $SO(n)$ on S^m, D^{m+1}, R^m, P^m, $m < 2n - 1$, $n \geqq 11$. Princeton Thesis, 1964.

KERVAIRE, MICHEL A.
A manifold which does not admit any differentiable structure. Comment. Math. Helv. 34 (1960), 257—270.

KINOSHITA, S.
On quasi translations in 3-space. Topology of 3-manifolds and related topics. Proc. of The University of Georgia Institute (1961), 223—226. Prentice-Hall, Inc., Englewood Cliffs, N. J., 1962.

KISTER, J. M.
Examples of periodic maps on euclidean spaces without fixed points. Bull. Amer. Math. Soc. (5) 67 (1961), 461—474.

A theorem on infinite regular neighborhoods and an application to periodic maps on E^n. Topology of 3-manifolds and related topics (Proc. The Univ. of Georgia Institute, 1961), 221—222. Prentice-Hall, Englewood Cliffs, N. J., 1962.

Differentiable periodic actions on E^8 without fixed points. Amer. J. Math. (2) 85 (1963), 316—319.

KISTER, J. M. and MANN, L. N.
Equivariant imbeddings of compact abelian Lie groups of transformations. Math. Annalen 148 (1962), 89—93.

Isotropy structure of compact Lie groups on complexes. Mich. Math. J. 9 (1962), 93—96.

KOBAYASHI, S.
Fixed points of isometries. Nagoya Math. J. 13 (1958), 63—68.

KWUN, KYUNG WHAN
An involution of the N-cell. To appear.

LIVESAY, G. R.
On maps of the three-sphere into the plane. Mich. Math. J. 4 (1957), 157—159.

Fixed-point-free involutions on the 3-sphere. Topology of 3-manifolds and related topics (Proc. The Univ. of Georgia Institute, 1961), 220. Prentice-Hall, Englewood Cliffs, N. J., 1962.

Involutions with two fixed points on the three-sphere. Ann. of Math. (3) 78 (1963), 582—593.

Involutions of the three-sphere with a circle of fixed points. To appear.

McMILLAN, D. R., JR.
Cartesian products of contractible open manifolds. Bull. Amer. Math. Soc. (5) 67 (1961), 510—514.

Some contractible open 3-manifolds. Trans. Amer. Math. Soc. (2) 102 (1962), 373—382.

McMILLAN, D. R., JR. and ZEEMAN, E. C.
On contractible open manifolds. Proc. Cambridge Phil. Soc. (2) 58 (1962), 221—224.

MANN, L. N.
Compact abelian transformation groups. Trans. of the Amer. Math. Soc. (1) 99 (1961), 41—59.

Finite orbit structure on locally compact manifolds. Mich. Math. J. 9 (1962), 87—92.

MANN, L. N. and SU, J. C.
Actions of elementary P-groups on manifolds. Trans. of the Amer. Math. Soc. (1) 106 (1963), 115—126.

MAZUR, BARRY
A note on some contractible 4-manifolds. Ann. of Math. (1) 73 (1961), 221—228.
Symmetric homology spheres. Ill. Jour. of Math. (2) 6 (1962), 245—250.
MILNOR, JOHN W.
On manifolds homeomorphic to the 7-sphere. Ann. of Math. (2) 64 (1956), 399—405.
Groups which act on S^n without fixed points. Amer. J. Math. 79 (1957), 623—630.
Remarks concerning spin manifolds. To appear.
MOISE, EDWIN
Periodic homeomorphisms of the 3-sphere. Ill. Jour. of Math. (2) 6 (1962), 206—225.
MONTGOMERY, DEANE
Finite dimensionality of certain transformation groups. Ill. J. Math. 1 (1957), 28—35.
Groups on R^n or S^n. Mich. Math. J. 6 (1959), 123—130.
MONTGOMERY, DEANE and MOSTOW, G. D.
Toroid transformation groups on euclidean space. Ill. J. Math. 2 (1958), 459—481.
MONTGOMERY, DEANE and SAMELSON, HANS
Transformation groups of spheres. Ann. of Math. (2) 44 (1943), 454—470.
Groups transitive on the n-dimensional torus. Bull. Amer. Math. Soc. 49 (1943), 455—456.
A theorem on fixed points of involutions in S^3. Canad. J. Math. 7 (1955), 208—220.
Examples of differentiable group actions on spheres. Proc. of the Nat. Acad. Sci. (8) 47 (1961), 1202—1205.
On the action of $SO(3)$ on S^n. Pacific Jour. Math. (2) 12 (1962), 649—659.
MONTGOMERY, DEANE; SAMELSON, H.; and YANG, C. T.
Exceptional orbits of highest dimension. Ann. of Math. (2) 64 (1956), 131—141.
Groups on E^n with $(n-2)$-dimensional orbits. Proc. Amer. Math. Soc. 7 (1956), 719—728.
MONTGOMERY, DEANE; SAMELSON, H.; and ZIPPIN, L.
Singular points of a compact transformation group. Ann. of Math. (2) 63 (1956), 1—9.
MONTGOMERY, DEANE and YANG, C. T.
The existence of a slice. Ann. of Math. (2) 65 (1957), 108—116.
Orbits of highest dimension. Trans. Amer. Math. Soc. 87 (1958), 284—293.
MONTGOMERY, D. and YANG, C. T.
Groups on S^n with principal orbits of dimension $n-3$. Ill. Jour. of Math. (4) 4 (1960), 507—517.
Groups on S^n with principal orbits of dimension $n-3$, II. Ill. Jour. of Math. (2) 5 (1961), 206—211.
A theorem on the action of $SO(3)$. Pacific Jour. of Math. (4) 12 (1962), 1385—1400.
MOSTERT, PAUL S.
Local cross sections in locally compact groups. Proc. of the Amer. Math. Soc. (4) 4 (1953), 645—649.
Sections in principal fibre spaces. Duke Math. J. (1) 23 (1956), 57—72.
Reasonable topologies for homeomorphism groups. Proc. Amer. Math. Soc. 12 (1961), 598—602.
MOSTOW, GEORGE DANIEL
Factor spaces of solvable groups. Ann. of Math. (2) 60 (1954), 1—27.
On covariant fiberings of Klein spaces. Amer. J. Math. 77 (1955), 247—278.

Equivariant embeddings in Euclidean space. Ann. of Math. (2) 65 (1957), 432—446.

On a conjecture of Montgomery. Ann. of Math. (2) 65 (1957), 513—516.

On the fundamental group of homogeneous space. Ann. of Math. (2) 66 (1957), 249—255.

Compact transformation groups of maximum rank. Bull. Soc. Math. Belg. 11 (1959), 3—8.

Cohomology of topological groups and solvmanifolds. Ann. of Math. (2) 73 (1961), 20—48.

NOMIZU, K.

Lie groups and differential geometry. Publications of the Math. Soc. of Japan, 1956.

PALAIS, RICHARD S.

Imbedding of compact, differentiable transformation groups in orthogonal representations. J. Math. Mech. 6 (1957), 673—678.

On the differentiability of isometries. Proc. Amer. Math. Soc. 8 (1957), 805—807.

A global formulation of the Lie theory of transformation groups. Mem. Amer. Math. Soc. 22 (1957), iii + 123 pp.

A covering homotopy theorem and the classification of G-spaces. Proc. Nat. Acad. Sci. U.S.A. 45 (1959), 857—859.

On the existence of slices for actions of non-compact Lie groups. Ann. of Math. (2) 73 (1961), 295—323.

Equivalence of nearby differentiable actions of a compact group. Bull. Amer. Math. Soc. 67 (1961), 362—364.

PALAIS, RICHARD S. and RICHARDSON, R. W., JR.

Uncountably many inequivalent analytic actions of a compact group on R^n. Proc. Amer. Math. Soc. 14 (1963), 374—377.

PALAIS, RICHARD S. and STEWART, THOMAS E.

Deformations of compact differentiable transformation groups. Amer. J. Math. 82 (1960), 935—937.

The cohomology of differentiable transformation groups. Amer. J. Math. 83 (1961), 623—644.

POÉNARU, VALENTIN

Les decomposition de l'hypercube en produit topologique. Bull. Soc. Math. France 88 (1960), 113—129.

RAYMOND, FRANK

The orbit spaces of totally disconnected groups of transformations on manifolds. Proc. Amer. Math. Soc. 12 (1961), 1—7.

RAYMOND, FRANK and WILLIAMS, R. F.

Examples of p-adic transformation groups. Bull. Amer. Math. Soc. 66 (1960), 392—394.

RICHARDSON, R. W., JR.

Actions of the rotation group on the 5-sphere. Ann. of Math. (2) 74 (1961), 414—423.

Groups acting on the r-sphere. Ill. Jour. of Math. (3) 5 (1961), 474—485.

ROSEN, RONALD H.

E^4 is the cartesian product of a totally non-euclidean space and E^1. Ann. of Math. (2) 73 (1961), 349—361.

Examples of non-orthogonal involutions of euclidean spaces. Ann. of Math. (3) 78 (1963), 560—566.

SMALE, STEPHEN
Generalized Poincaré's conjecture in dimensions greater than four. Ann. of Math. (2) 74 (1961), 391—406.
Differentiable and combinatorial structures on manifolds. Ann. of Math. (3) 74 (1961), 498—502.

SMITH, P. A.
Orbit spaces of abelian p-groups. Proc. Nat. Acad. Sci. U.S.A. 45 (1959), 1772—1775.
New results and old problems in finite transformation groups. Bull. Amer. Math. Soc. 66 (1960), 401—415.
The cohomology of certain orbit spaces. Bull. Amer. Math. Soc. (4) 69 (1963), 563—568.

STEWART, T. E.
On R-equivalent spaces. Nederl. Akad. Wetensch. Proc. Ser. A. 61 = Indag. Math. 20 (1958), 460—462.
Uniqueness of the topology in certain compact groups. Trans. Amer. Math. Soc. 97 (1960), 487—494.
Lifting the action of a group in a fibre bundle. Bull. Amer. Math. Soc. 66 (1960), 129—132.
On groups of diffeomorphisms. Proc. Amer. Math. Soc. 11 (1960), 559—563.
Fixed point sets and equivalence of differentiable transformation groups. Comm. Math. Helv. 38 (1963), 6—13.

WHITEHEAD, J. H. C.
On normalizators of transformation groups. J. London Math. Soc. 27 (1952), 374—379.
On involutions of spheres. Ann. of Math. (2) 66 (1957), 27—29.

YANG, CHUNG-TAO
On a problem of Montgomery. Proc. Amer. Math. Soc. 8 (1957), 255—257.
Transformation groups on a homological manifold. Trans. Amer. Math. Soc. 87 (1958), 261—283.
p-adic transformation groups. Mich. Math. J. 7 (1960), 201—218.
The triangulability of the orbit space of a differentiable transformation group. Bull. Amer. Math. Soc. (3) 69 (1963), 405—408.

INDEX